Positive Unlabeled Learning

Synthesis Lectures on Artificial Intelligence and Machine Learning

Editors

Ronald Brachman, *Jacobs Technion-Cornell Institute at Cornell Tech*
Francesca Rossi, *IBM Research AI*
Peter Stone, *University of Texas at Austin*

Series Page

Positive Unlabeled Learning

Kristen Jaskie and Andreas Spanias

ISBN: 978-3-031-79173-4   paperback
ISBN: 978-3-031-79178-9   PDF
ISBN: 978-3-031-79183-3   hardcover

DOI 10.1007/978-3-031-79178-9

A Publication in the Springer series
SYNTHESIS LECTURES ON ARTIFICIAL INTELLIGENCE AND MACHINE LEARNING

Lecture #51
Series Editors: Ronald Brachman, *Jacobs Technion-Cornell Institute at Cornell Tech*
                Francesca Rossi, *IBM Research AI*
                Peter Stone, *University of Texas at Austin*
Series ISSN
Print 1939-4608   Electronic 1939-4616

Positive Unlabeled Learning

Kristen Jaskie
Arizona State University

Andreas Spanias
Arizona State University

SYNTHESIS LECTURES ON ARTIFICIAL INTELLIGENCE AND MACHINE LEARNING #51

ABSTRACT

Machine learning and artificial intelligence (AI) are powerful tools that create predictive models, extract information, and help make complex decisions. They do this by examining an enormous quantity of labeled training data to find patterns too complex for human observation. However, in many real-world applications, well-labeled data can be difficult, expensive, or even impossible to obtain. In some cases, such as when identifying rare objects like new archeological sites or secret enemy military facilities in satellite images, acquiring labels could require months of trained human observers at incredible expense. Other times, as when attempting to predict disease infection during a pandemic such as COVID-19, reliable true labels may be nearly impossible to obtain early on due to lack of testing equipment or other factors. In that scenario, identifying even a small amount of truly negative data may be impossible due to the high false negative rate of available tests. In such problems, it is possible to label a small subset of data as belonging to the class of interest though it is impractical to manually label all data not of interest. We are left with a small set of positive labeled data and a large set of unknown and unlabeled data. Readers will explore this Positive and Unlabeled learning (PU learning) problem in depth.

The book rigorously defines the PU learning problem, discusses several common assumptions that are frequently made about the problem and their implications, and considers how to evaluate solutions for this problem before describing several of the most popular algorithms to solve this problem. It explores several uses for PU learning including applications in biological/medical, business, security, and signal processing. This book also provides high-level summaries of several related learning problems such as one-class classification, anomaly detection, and noisy learning and their relation to PU learning.

KEYWORDS

machine learning, artificial intelligence, neural networks, semi-supervised, signal processing, positive unlabeled learning, big data, detection, classification

Contents

Preface

Positive Unlabeled learning (PU learning) is a semi-supervised learning method that classifies and labels data in one of the most difficult of all possible classification scenarios. Other semi-supervised learning methods use a small amount of labeled data from *each* class and make use of additional unlabeled data to improve classification results. PU learning handles the situation when only some data from *one class* is known and *everything else* is unlabeled. No labeled data exists from the other set and often few labeled data exists at all. And yet, algorithms exist that can solve this problem and label the large amounts of unlabeled data effectively!

This book was motivated in part by a series of PU learning seminars at the SenSIP center at Arizona State University which led to increased interest and further research on the subject. This led to both a short survey paper and an expanded tutorial on the subject presented at IEEE IISA—the 10th International Conference on Information, Intelligence, Systems and Applications in Greece in 2019, with publication of the MLR algorithm later in the year at the IEEE Asilomar Conference on Systems, Signals, and Computers. Further research into PU applications and applicability led to the applied PU use case in photovoltaic solar fault classification published in Jaskie, Martin, and Spanias [2021]. This focus in applications and use cases for PU learning appears to be relatively unique in the field and provided impetus for an expanded chapter at the end of this text.

The primary objective of this book is to introduce and define the PU learning problem, provide a broad survey and literature review of both its algorithmic development and current research, and introduce some of the many and varied applications and use cases to which it is applied. It is intended for a relatively broad spectrum of readers, including graduate students, faculty, and industry/government practitioners. The book assumes a moderate background in machine learning concepts and algorithms but provides enough material to catch up a newcomer to the field in all but the most complex algorithms, which are indicated as such and able to be skipped over without loss of a conceptual understanding.

Kristen Jaskie and Andreas Spanias
January 2022

Acknowledgments

The study was supported in part by various the NSF awards. Logistical support was provided by the ASU SenSIP center. We cite specifically NSF awards 1525716, 155040, 1659871, 1854273, 1953745, and 2019068 for support of the authors at various stages of this project. The authors would also like to thank SenSIP industry members NXP, Raytheon, and especially the Prime Solutions Group (PSG) for support in this and associated research activities in machine learning.

Kristen Jaskie and Andreas Spanias
January 2022

CHAPTER 1

Introduction

Although the foundations of artificial intelligence (AI), and more specifically machine learning (ML), have been around for several decades, their utility in every day applications has increased considerably in the last ten years. This is due in part to the emergence of new applications involving "big" data sets and advancements in computing, wireless communications, and network sensing to name a few.

As capabilities in AI and ML grow, the potential of analyzing big data sets with precision and confidence is also improving. We have recently witnessed not only massive ML research endeavors but also well-proliferated ML-enabled products, such as Internet of Things (IoT) smart speakers, and intelligent smartphone applications (apps) addressing health, sustainability, business, energy, security, and entertainment. Regardless of the success in launching applications, education and workforce creation in ML is both a challenge and an opportunity. This is because the AI and ML theory of algorithms involves advanced mathematics and statistics that are typically part of a graduate curriculum. Nevertheless, several books and tools have emerged to introduce the novice to ML. In this book, we review basic ML methods, but we focus specifically in introducing and studying the relatively new field of positive unlabeled (PU) learning.

1.1   WHAT IS POSITIVE UNLABELED LEARNING?

We live in the era of "Big Data" where huge amounts of data are collected, but little is *actually known* or *understood*. For example, the human genome project completed mapping human DNA in 2003 and we now have a list of the approximately 20,500 genes in the human genome. We have vast amounts of data, however we need to extract useful information for the problem at hand. The process of learning from these long data sequences is called "knowledge extraction." We would like to be able to identify which genes may cause or influence diseases that are known to have a strong genetic component, one example being Alzheimer's disease. Many genes have so far been found that influence Alzheimer's disease, but have we found all of them?

To solve these problems and extract this information, we need to create a mathematical model of the data. Traditional software development requires the programmer to define the rules and parameters of a model themselves. The programmer needs to understand all the rules, components, and assumptions involved, and then write a detailed program to create a precise software model. Any assumptions and initial conditions that the programmer leaves out or does not incorporate correctly, negatively influence the final model. Machine learning (ML) works differently. In ML, the model is influenced more from the data itself than from the programmer.

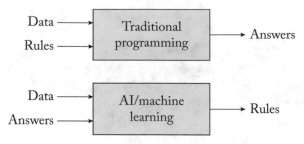

Figure 1.1: Machine learning requires a new way of thinking compared to traditional programing.

The programmer defines only the most basic of model parameters and the program "learns" the rest from the data. This process can often find far subtler patterns and correlations than are possible with other, more traditional modeling methods (Figure 1.1).

There are several different types of ML algorithms, but they are generally divided and categorized based on how much and what type of information they use. The three commonly mentioned main categories are supervised learning, where all the answers, or labels, are known ahead of time, unsupervised learning, where no answers are known, and reinforcement learning, where an environment exists to be explored to obtain the answers. In between supervised and unsupervised learning is a less commonly described hybrid of interest in this book, known as semi-supervised learning (also called semi-supervised classification)—where some answers are known, and others are not known. These ML categories are discussed in more depth in the next section.

In the human genome example above, our objective is to identify genes that may influence a specified disease such as Alzheimer's. To do this, we look at the entire human genome along with the handful of genes that are already known to influence the disease. We consider any genes that are not known to influence the disease to have *unknown* influence rather than no influence because in truth, the subject is so complicated that genes for one thing can indeed have unknown influence over other things. It is nearly impossible to be 100% sure that a given gene does NOT in some way subtly influence some disease. This scenario is one example of a Positive Unlabeled learning problem (PU learning)—some positive and a large amount of unknown data is available. Our treatment of PU learning will include further problem motivation, careful problem formulation, solutions for solving and evaluating this problem, important applications, and an extensive bibliography to assist further research in this area.

PU scenarios occur surprisingly often in the real world. Some data from the positive class (the class of interest) is available, while no guaranteed data from the negative class is known with any certainty resulting in a positive and unlabeled (PU) dataset. Recommender systems, such as Netflix and Amazon Prime, recommend shows and movies that you may like to see. Shows that you have watched before, and thus clearly liked, can be considered as elements of the positive class, and all shows that you have not watched are considered unknown—you may or may not like them. The recommender's job is to classify the unwatched shows and movies as something you will like or dislike. This is also used in personalized advertisements. As you browse the web, items that you search for, click on, or purchase show interest. These become the positive samples. All others are considered unknown. Advertisers want to identify which other products or services you may like—a perfect PU application. It is not just recommender systems however, applications include identifying military vehicles or archeological ruins in satellite pictures, detecting early cancer cases from medical records, identifying credit card fraud or manufacturing faults, and even predicting how members of parliament will vote on an issue are all areas of PU research. Chapter 5 looks at these various applications in much more depth.

1.2   A BRIEF OVERVIEW OF MACHINE LEARNING

While it is out of the scope of this book to provide a thorough or exhaustive introduction to machine learning, a general overview is useful for understanding how classification algorithms and the PU learning problem fit in the grand scheme of machine ML and AI. As mentioned above, ML algorithms are commonly divided into three broad categories—supervised learning, unsupervised learning, and reinforcement learning. The type of data that is available and the problem to be solved dictates which of these three general types is used. Many different problems can be solved using ML including classification, regression, clustering, compression, and data synthesis, among others. For the purposes of this book, we are most interested in the classification problem and will focus on it in the paragraphs below. Classification is the sorting of data into two or more classes or categories that are set by the user. When there are only two classes, the problem is called a binary classification problem and the two classes are often renamed the positive and negative classes. The positive class is generally the smaller of the two or the one of the most interest if balanced. Figure 1.8 at the end of this section provides a snapshot overview of some of the more common ML categories and problems and where PU learning fits within the overall scheme of things.

Probably the most common type of ML in use today is supervised learning. Supervised learning requires complete labels to be available for the dataset that is to be modeled. Labels can be thought of as the "answer" that your algorithm will be solving for when given new, never seen, data. For example, in Figure 1.2, the dataset to be modeled is binary and includes images from two classes—cats and dogs. As important, labels are provided that specify which images are of dogs and which are of cats. Given this labeled dataset, a supervised learning algorithm can be used to learn a classification model that can differentiate between images of cats and dogs. Then, given a new image with unknown label, this model can predict whether the new image is of a cat or a dog with some degree of certainty. This is an example of supervised classification.

Unsupervised learning takes data with no labels as input. Unsupervised learning algorithms generally attempt to find patterns in input data that allow it to group, or cluster, these inputs together based on underlying similarity of the data. Other types of unsupervised learning

Figure 1.2: Example of a supervised learning classification model with training images on the left and a new image, on the bottom right, to be classified by the trained model.

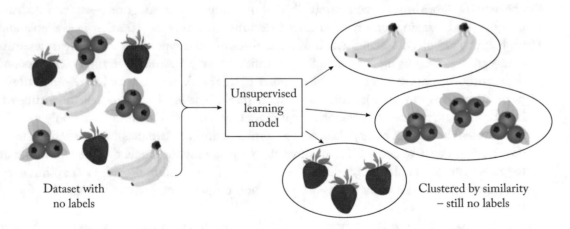

Figure 1.3: Example of an unsupervised learning model when no labels are present to train the model.

algorithms are used for dimensionality reduction, like compression. In Figure 1.3, the data to be clustered is composed of different types of fruit. Notice that none of the fruit is labeled with its name or other identifying information. This fruit data could be images of the fruit, or it could be composed only of its nutritional information, or size and weight, or any other measured characteristics or combinations thereof. The unsupervised learning algorithm learns a model from this unlabeled data that clusters similar items together. The algorithm has no way of knowing that the long yellow fruit is a banana as that label was never provided, but it can combine all long yellow fruit into a single group, or cluster. New data samples can also be sent to their appropriate groups/clusters using this model. This is commonly referred to as clustering, though it can be thought of as a form of unsupervised classification, where the identification of each class remains unknown.

In between supervised and unsupervised learning lies the very important, yet less commonly discussed, category of semi-supervised learning. Semi-supervised learning is important because it occurs so often in real-world data. It is often quite difficult, expensive, or even impossible to obtain labels for *all* data samples in the real world. Semi-supervised learning algorithms deal with the classification scenario where some, but not all the data is labeled. It has been shown that *using* unlabeled data with labeled data rather than discarding them can generally improve semi-supervised models [De Comité et al., 1999]. Traditional semi-supervised learning uses some labeled data from all classes along with whatever unknown/unlabeled data is available to improve the model. The PU learning problem is a binary classification problem with labeled data available from only one class, rather than both as in traditional semi-supervised learning. This difference is illustrated in Figure 1.4. Traditional semi-supervised learning is an easier problem than positive unlabeled learning as more information is available. Chapelle, Schölkopf, and Zien [2006] is a great reference book for the general semi-supervised learning problem.

It is surprisingly common to have some data with known positive labels and the rest unknown, meaning no negative labels are available even though the unknown set is composed of both positive and negative data. This is the PU learning problem that is the focus of this book. PU learning scenarios occur quite frequently, though they are not always recognized as such due to the general focus on supervised classification over lesser-known semi-supervised algorithms. This is a mistake, as we will show in later chapters that using a supervised learning algorithm on a PU dataset can result in a very poor model. PU learning is appropriate whenever some data with a property of interest is available, while the rest is unknown. Examples of this include fraud detection, object identification, medical detection, and many more.

An illustration of this is the disease gene example given in the previous section is shown in Figure 1.5. We know of *some* genes that influence a disease such as Alzheimer's, but it is extremely expensive, if not impossible, to state whether a *particular* gene has absolutely no influence on a disease. In this case, a PU learning model is created which assigns labels to all the unknown samples.

The final general category of machine learning is reinforcement learning, shown in Figure 1.6. Reinforcement learning is not commonly used with classification algorithms, though some PU algorithms try to boost the importance of the few known positive samples using reinforcement techniques. General reinforcement learning is very similar to how young children learn about their world. If a child does something good such as eating a strawberry, they get positive feedback from their senses in the form of a good taste. When the child does something bad such as eating or trying to eat a houseplant they receive negative feedback. In ML, reinforcement learning usually involves one or more "agents" (these can be anything from software

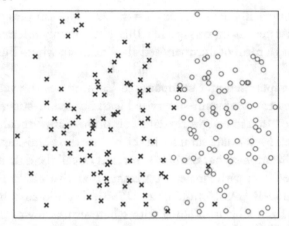

(a) Fully labeled dataset. What is available for supervised learning.

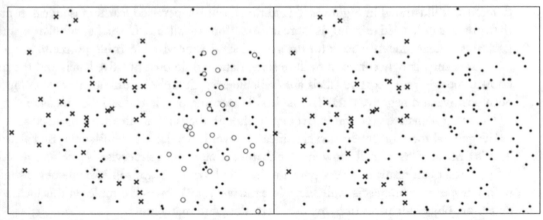

(b) Semi-supervised dataset with some labeled datapoints from both classes available

(c) Positive unlabeled dataset with labeled datapoints from only one class available

Figure 1.4: The same data as a supervised dataset (a), traditional semi-supervised dataset (b), and a PU dataset (c).

algorithms to physical robots) that perform actions in some sort of an environment—simulated or physical. As they interact with the environment or each other, they get positive or negative reinforcement, often via a loss function created by their designers. This is how DeepMind's AlphaZero algorithm was able to teach itself how to play Chess and Go better than any human in only a couple of hours [Silver et al., 2018].

One type of reinforcement learning that has been used in some recent PU learning algorithms are *Generative Adversarial Networks*, more commonly called GANs. A GAN is used to

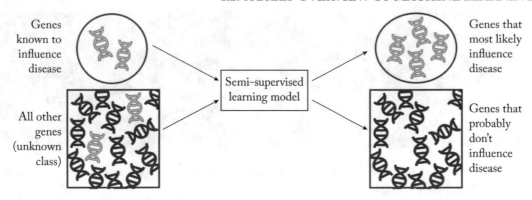

Figure 1.5: Example of a positive unlabeled learning model where some genes of interest are labeled, and all others are unlabeled.

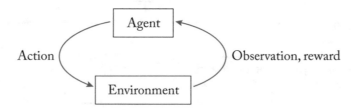

Figure 1.6: A high level illustration of reinforcement learning showing how the agent and the environment interact with one another. Reinforcement learning is not generally used to solve classification or regression problems.

generate new, realistic images or data that can be indistinguishable from its training data. Deep fake images are an example of what can be done with a GAN and can be seen in Figure 1.7. In some newer PU algorithms, GANs are used to try to create additional positive or positive and negative data samples to improve learning. Other PU algorithms use a simpler reinforcement technique called boosting to improve their algorithms. These will be discussed in more detail in Chapter 4.

The brief amount of ML overview we've been able to provide in this section should give you an idea of how PU learning fits into the grand scheme of ML categories and problems. A simplification of this can be visualized in Figure 1.8. What is surprising about PU learning is its high potential in an array of important real-world applications. Yet little is relatively known about PU learning and its many applications. In this book, we plan to focus on this important problem and the algorithms and software that exist to address it.

Figure 1.7: **Example of fake images generated by a GAN** at http://www.whichfaceisreal.com, [West and Bergstrom].

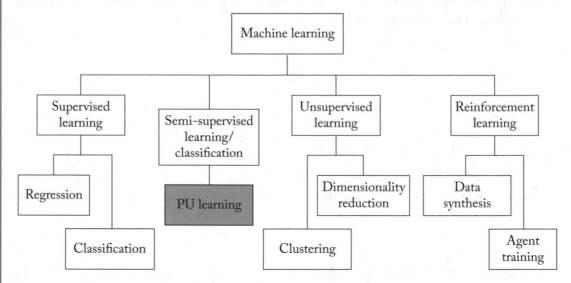

Figure 1.8: Taxonomy of ML algorithms and PU learning. PU learning is a semi-supervised learning algorithm that can learn a classification model from only some positive and accompanying unlabeled data.

1.3 PROBABILITY THEORY AND ITS RELATION TO CLASSIFICATION

As we mentioned in the previous section, positive unlabeled learning is a type of semi-supervised classification algorithm. There are many ways to define classification algorithms—they can be described by the mathematical functions that are learned, the complex graphs that are created and traversed, the hyperplanes that are found, and the structure of any neural networks that are used.

$$\underset{\text{probability}}{\underset{\text{Posterior}}{}} p(y=1|x) = \frac{\overset{\text{Likelihood}}{p(x|y=1)} \cdot \overset{\underset{\text{(class prior)}}{\text{Prior probability}}}{p(y=1)}}{\underset{\underset{\text{probability}}{\text{Marginal}}}{p(x)}}$$

Figure 1.9: Labeled probabilistic components and their relationship in Bayes' Theorem.

The output of any classification algorithm is a one-to-one function that, when given an input datapoint, provides a prediction of the class label for that datapoint. Some types of algorithms such as logistic regression and neural network classifiers produce a probabilistic output that is transformed into a classification by defining a class threshold, most commonly 50%. That is, if the algorithm predicts that a datapoint x belongs to the positive class with a probability of 50% or more, we classify x as belonging to the positive class. Other types of algorithm output, such as support vector machines (SVMs) provide the classification as the discrete final decision only. Probability calibration techniques such as Platt scaling [Platt, 1999, Ruping, 2006] or isotonic regression [Zadrozny and Elkan, 2002] are sometimes used to estimate the associated probabilities. Because of this, for simplicity and the purposes of this book, we will treat all classifiers as being probabilistic in nature.

In its simplest form, a binary classifier can be thought of as a function that provides an estimate of the conditional posterior probability that a data sample belongs to the positive class given its feature values. We will use standard notation and describe a data sample by its feature vector $x \in \mathbb{R}^n$. The class label $y \in \{0, 1\}$ is used to label the data sample as positive ($y = 1$) or negative ($y = 0$). Using this notation, a classifier can be written as a function $f : \mathbb{R}^n \rightarrow [0, 1]$.

$$f(x) = p(y = 1|x). \tag{1.1}$$

This posterior probability $p(y = 1|x)$ is related to the prior probability $p(y = 1)$, the likelihood $p(x|y = 1)$, and the marginal probability of observing x, $p(x)$, via Bayes' theorem as shown in Figure 1.9. The marginal probability $p(x)$ is typically ignored as a constant, as the features and data x are known to exist and do not depend on the value of y.

The *class prior* probability $p(y = 1)$ is sometimes referred to as the *a priori* probability and is often represented using the symbol π. While these terms are often used interchangeably, the *a priori* is more of a philosophical notion that represents belief in the prior probability before data collection. Looking at a single die, you might "a priori" judge that it is even and fair, meaning that you think that the probability of rolling a 1 is 1/6. On the other hand, the prior probability of rolling a 1 is determined only after rolling the die many times and finding that in fact, it occurs once out of every six rolls on average. One is a belief, the other is evidence based. In supervised classification, the class prior is something that can be calculated—in this case the percentage of

Table 1.1: Probability terms and formulae used in this book

Term	Rule	Description
Probability range	$0 \leq (y = 1) \leq 1$	All probability values are between zero and one.
Posterior probability	$p(y = 1 \mid x)$	The probability that datapoint x is positive, given its feature vector.
Likelihood	$p(x \mid y = 1)$	This describes how likely the datapoint is to exist given that it is positive. If x is in the center of the positive distribution, then it is likely. If it is an outlier, it is less likely.
Prior probability	$p(y = 1)$	The probability that a sample is positive given no other information. Abbreviated π.
Marginal probability	$p(x)$	This is the marginal probability of observing datapoint x. Because x was observed, this is commonly dismissed as a constant.
Bayes' rule	$p(A \mid B) = \dfrac{p(B \mid A) \cdot p(A)}{p(B)}$	This is the generic form of Bayes' law, also called Bayes' rule or Bayes' theorem. Describes the relationship between the posterior probability, likelihood, prior, and marginal probabilities.
Bayes' rule for classification	$p(y = 1 \mid x) = \dfrac{p(x \mid y = 1) \cdot p(y = 1)}{p(x)}$	The application of Bayes' Rule or Bayes' Theorem to a probabilistic binary classifier. Also provided in Figure 1.9.

positive data samples out of all data samples. In semi-supervised classification (including PU learning), a class prior exists, but it may need to be estimated as not all labels are known.

1.4   INTENDED AUDIENCE

This book is designed for graduate students, faculty, practitioners, and ML researchers interested in the PU learning problem. PU learning is described, theory and variants are discussed, and algorithms are presented in detail. Applications and examples are discussed at length and a comprehensive bibliography is provided which will be useful for launching new research in this area. An intermediate level of linear algebra and programming experience is assumed. Some algorithm descriptions require a background in ML to fully understand.

1.5   BOOK ORGANIZATION

This book attempts to cover several topics related to PU learning. Although the manuscript is designed for beginning-to-end reading, readers may choose to read self-contained chapters and sections as needed. This section will provide a short description of the chapters in this book and their purpose. At the end of the book, we provide an extensive Bibliography and Appendices with symbol definitions and additional resources.

Chapter 2 describes the PU learning problem in detail, providing real-world examples as motivation. The PU learning problem is formalized and a description and brief explanation of all variables and commonly used terms is provided. PU data collection techniques and common assumptions are described in depth. Common variants of the PU learning problem are briefly introduced.

Chapter 3 explains the unique requirements for evaluating PU learning algorithms and solutions. Because of the small amount of labeled data available in PU learning, specialized evaluation techniques are required.

Chapter 4 provides several algorithms for the PU learning problem that have been proposed over the last 20 years. A survey of solutions is given with selected high-performing solutions described in more detail.

Chapter 5 describes several real-world applications for PU learning. These applications are taken from the literature as well as new proposed applications selected due to their inherent PU nature.

Chapter 6 provides a brief summary and conclusion.

CHAPTER 2

Problem Definition

In this chapter, we present the Positive and Unlabeled (PU) learning problem in detail. In addition to notation and a formal statement of the problem, we explore fundamental factors that influence PU models such as how the source data for the model was acquired and what assumptions are made about that data. We end the chapter with a brief description of two closely related learning problems, one class classification (OCC) and noisy learning.

2.1   MOTIVATION

To motivate PU learning, we start by examining a typical supervised classification situation and discuss potential problems and pitfalls that could be corrected using a PU learning approach.

In Table 2.1, we see a subset of a well-known classification dataset known as the Pima Indians Diabetes Dataset [National Institute of Diabetes and Digestive and Kidney Diseases, 2016]. Each *row* in the table describes a person, or "data sample," for our model. The data features are shown in the first eight *columns*. These features describe the number of pregnancies and medical test results for each patient. The binary outcome (or label)—whether the patient developed diabetes within the following five years—is shown in the rightmost column. The classification goal is to learn a model that can predict whether a new patient with their own medical data (data features) will develop diabetes in the next five years or not. This is a standard supervised classification problem.

While this well-labeled supervised dataset provides information to help identify diabetes before it occurs, it is extremely limited and biased toward the fewer than 20,000 Pima in the southwest United States and is thus not generalizable to a global population. Worldwide, over 422 million adults were known to have diabetes in 2014 and approximately 1.6 million deaths directly attributable to diabetes in 2016 according to the World Health Organization [Diabetes, 2018]. To effectively learn an unbiased supervised classification model, a large, multi-year study would be required to collect sufficient labeled data. The cost for such a study would be enormous.

Even with such staggering numbers of diabetics worldwide, studies suggest that diabetes frequently goes undiagnosed for many years [Geiss et al., 2018]. It would be extremely valuable to be able to inexpensively identify individuals with currently undiagnosed diabetes so they could get medical attention before the disease gets out of control. This could save hundreds of thousands of lives annually.

Rather than conducting expensive new tests, it would be relatively inexpensive to look at *existing* medical records and previously collected metabolic panels for many people around

Table 2.1: Subset from the Pima Indians Diabetes Dataset [National Institute of Diabetes and Digestive and Kidney Diseases, 2016]

Data Features								Labels
Number of pregnancies	Plasma glucose	Diastolic blood pressure	Triceps skinfold thickness	2-hour serum insulin	Body mass index	Diabetes pedigree function	Age	Diabetes within 5 years
6	148	72	35	0	33.6	0.627	50	1
1	85	66	29	0	26.6	0.351	31	0
1	89	66	23	94	28.1	0.167	21	0
0	137	40	35	168	43.0	2.288	33	1
3	126	88	41	235	39.3	0.704	27	0
⋮	⋮	⋮	⋮	⋮	⋮	⋮	⋮	⋮

the world—a much larger and less biased sample than looking just at one small indigenous population. Those persons with a known diabetes diagnosis could be labeled as having diabetes, while those persons who have not been diagnosed with diabetes could be labeled as having *unknown* status, rather than *negative*. We now have a PU learning problem, that can be used with currently available information at a much lower cost.

In this, as in many real-world problems, it can be difficult, expensive, or even impossible to gather enough labeled data for effective supervised training. In some cases, such as when identifying rare objects like new archeological sites or secret enemy military facilities in satellite images, acquiring labels could require months of trained human observers at incredible expense. Other times, as when attempting to predict disease infection during a pandemic such as Covid-19, reliable true labels may be nearly impossible to obtain early on due to lack of testing equipment or other factors. In that scenario, identifying even a small number of truly negative datapoints can be impossible due to the high false negative rate of available tests. Companies such as Netflix and Hulu want to recommend movies that you may like to watch, but without spending time rating every movie you've seen before, very few if any negative labels may be available. In scenarios such as these, supervised learning is impossible or impractical, while PU learning allows effective classification. Later in the book, in Chapter 5, we explore many varied PU applications.

2.2   PROBLEM DESCRIPTION AND FORMULATION

In this section, we formalize the positive unlabeled learning problem (PU learning) and describe the notation that will be used in this book.

The general classification problem is an important and common task in machine learning (ML) that is used in a wide variety of fields from the biological and biomedical, to business ventures, signal processing, security applications, and many more. Classification problems are defined as either binary or multi-class. In binary classification, exactly two classes of data are available and must be separated from one another. These two classes are typically referred to as the positive and negative classes. Examples of binary classification include identifying images of cats vs. dogs or fraudulent credit card transactions vs. legitimate transactions. Multi-class classification defines the problem where three or more classes of data are to be differentiated. In some situations, multi-class classification can be transformed to a series of binary classifications—isolating each class from the rest, one at a time. When a single class is to be identified from a non-homogonous group of data composed of multiple classes, this is known as one-class or one-vs.-rest classification. One example is to identify images of a single species of animal from an image database of many types of animals.

While PU learning is often used in practice for both binary and one-class classification, the problem is generally formalized as a strictly binary problem. In this book we have followed this standard problem format and define PU learning as a binary classification problem. One-class classification (OCC) will be described separately in Section 2.4.1.

We begin formulating the problem by carefully defining the binary *supervised* classification problem so we can illustrate how the PU learning problem differs.

2.2.1   SUPERVISED BINARY CLASSIFICATION

In supervised binary classification, each data sample x, consisting of n features, is stored in a $1 \times n$ feature vector that corresponds to a single row in the data matrix X, shown in Figure 2.1. The label for data sample x is a binary value stored in binary variable $y \in \{0, 1\}$. When $y = 1$, x is said to belong to the positive class, and when $y = 0$, x belongs to the negative class. The $m \times n$ data matrix consisting of m total data samples x is denoted X, and the $n \times 1$ column vector storing the associated labels is y.

A traditional classification algorithm learns a model

$$f : \mathbb{R}^{1 \times n} \to [0, 1] \tag{2.1}$$

from the labeled training samples X and y such that a new data sample $\widehat{x}$, can be used as input to the trained model f to predict $\hat{y} : f(\widehat{x}) = \hat{y}$. This model can produce either binary output or a probabilistic real-valued output. As binary output can be considered a special case of a probabilistic output with only two values (zero and one), we will consider the output of our model, $f(x)$ or $\hat{y}$, to be probabilistic unless otherwise stated.

Given a data sample x, the model $f(x)$ should return the conditional probability that x belongs to the positive class ($y = 1$) given its feature vector:

$$f(x) = p(y = 1 | x). \tag{2.2}$$

In traditional supervised classification, many algorithms can be used to learn the model $f(x)$ including most commonly logistic regression, support vector machines (SVMs), random forest classifiers, and both shallow and deep neural networks (NNs) depending on the complexity

n feature values

Data Features								Labels
Number of pregnancies	Plasma glucose	Diastolic blood pressure	Triceps skinfold thickness	2-hour serum insulin	Body mass index	Diabetes pedigree function	Age	Diabetes within 5 years
6	148	72	35	0	33.6	0.627	50	1
1	85	66	29	0	26.6	0.351	31	0
1	89	66	23	94	28.1	0.167	21	0
0	137	40	35	168	43.0	2.288	33	1
3	126	88	41	235	39.3	0.704	27	0
⋮	⋮	⋮	⋮	⋮	⋮	⋮	⋮	⋮

m data samples x

X
$m \times n$

y
$m \times 1$

Figure 2.1: Classification notation used in this book. Each data sample x is represented as a $1 \times n$ horizontal feature vector in the data matrix X which contains all m data samples. True data labels for each data sample are stored in column vector y. Data sample x_i corresponds with label y_i.

of the problem. Where the probabilistic output of $f(x)$ reaches some threshold value (commonly 50%), it defines a classification boundary between the positive and negative classes, transforming the probabilistic output to a binary decision.

2.2.2   POSITIVE UNLABELED CLASSIFICATION

In PU learning, the class label y is not available for all training data. Instead, some samples from the positive class are known and labeled while the remaining samples, both positive and negative, are unknown and unlabeled. The underlying data distributions are unknown.

To handle this, a new random variable s is introduced that represents whether a sample is *labeled* or not. If a sample x is *labeled* positive, then its associated $s = 1$. If the sample is *unlabeled*, meaning it is unknown whether it is positive or negative, then $s = 0$. The PU problem can be stated formally using this notation as

$$p(s = 1|y = 0) = 0. \tag{2.3}$$

In PU learning, as in traditional classification, the goal is to learn a model $f(x) = p(y = 1|x)$ that can differentiate between the positive and negative classes. When $x \in \mathbb{R}^n$, this can be thought of as creating a decision boundary in n-dimensional space. This is a more difficult problem than traditional classification, as less information is available about these classes.

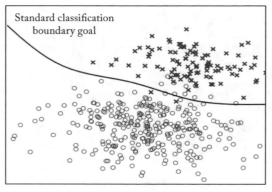
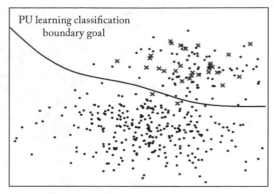

(a) Supervised classification learning problem
with all labels known

(b) Positive unlabeled learning problem with
only a percentage of known positive labels

Figure 2.2: Illustration of the positive unlabeled learning problem in two dimensions. The true class label is given by the shape and color shown in (a). This information would be available to a supervised learning algorithm. The information available to a PU learning algorithm is shown in (b) with the black dots representing unknown/unlabeled data. Notice the complete overlap between the known positive (red x's) and unlabeled data.

The variable s represents whether a data sample x is *labeled* or *unlabeled*, NOT whether it is *positive* or *negative*.

Figure 2.2 provides a two-dimensional illustration of the differences between traditional classification and PU learning. In this example, the feature vector for each sample contains the vertical and horizontal coordinates of each datapoint. The true positive and negative classes are shown by color and shape in Figure 2.2a while the PU learning scenario is shown in Figure 2.2b. Unknown samples in Figure 2.2b are marked with small black dots while the few labeled positives are marked by red x's. Despite missing substantial information in Figure 2.2b, our goal is the same—to learn the *same* decision boundary between the positive and negative classes.

What data is available?

- In supervised classification, data pairs $< x, y >$ are used for training.

- In PU classification, only $< x, s >$ is available. y is a latent variable to be predicted by the model $f(x) = p(y = 1|x)$.

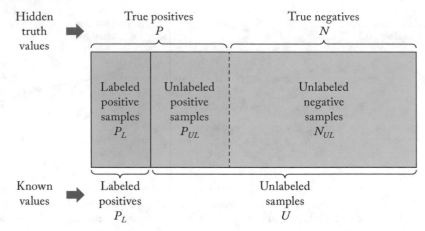

Figure 2.3: The PU learning problem can be formulized using sets. Here, true class values P and N are shown at the top but are unknown. Available information is shown at the bottom with P_L and U. The PU learning goal is to identify hidden sets P and N.

The positive and negative data are typically considered to be independent and identically distributed (i.i.d.) samples from the positive and negative data distributions, $p(x|y = 1)$ and $p(x|y = 0)$, respectively. This means that the distribution of x, $p(x)$, can be thought of as a two-component mixture model:

$$p(x) = p(y = 1)p(x|y = 1) + p(y = 0)p(x|y = 0). \qquad (2.4)$$

As the class prior $p(y = 1)$ is often represented using the pi symbol π, this can also be written as

$$p(x) = \pi \cdot p(x|y = 1) + (1 - \pi) \cdot p(x|y = 0). \qquad (2.5)$$

Set Notation

In addition to the probabilistic notation described above, PU learning and PU datasets are sometimes described using set notation. In this case, the data is described as belonging to sets of positive, negative, and unlabeled data referred to as P, N, and U, respectively. Using this notation, traditional classification can be thought of as identifying a decision boundary between sets P and N, where $X = P \cup N$. PU learning is given only the labeled positive set P_L, and the unlabeled set U, as shown in Figure 2.3. In this case, $X = P_L \cup U$ where the unlabeled set U is composed of both unlabeled positive samples P_{UL} and unlabeled negative elements $N_{UL} : U = P_{UL} \cup N_{UL}$. Notice that $P = P_{UL} + P_L$ and $N = N_{UL}$. Some papers in the literature use P to represent the known positive samples and Q for positive samples that are unlabeled, but to avoid confusion, in this book, we will use P_L and P_{UL}, respectively.

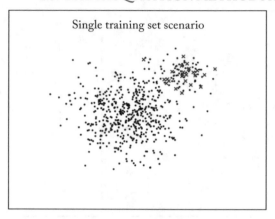

Figure 2.4: The single training set scenario with random sampling from a single dataset.

2.3   DATA ACQUISITION METHODS AND ASSUMPTIONS

The way data is acquired and labeled can have a significant impact on the effectiveness of PU learning algorithms. The collection and labeling mechanisms are often separated in the literature.

2.3.1   DATA ACQUISITION METHODS

Data acquisition or collection can be performed using either the *single training set* scenario, sometimes called the *censoring scenario* [Kato, Teshima, and Honda, 2019] or the *case-control scenario*.

Single Training Set Scenario

In the single training set scenario, all data, both labeled positive and unlabeled, are collected from the general data distribution $p(x)$ as shown in Figure 2.4. In this collection method, a positive data sample is labeled positive with some probability (discussed in Section 2.3.2) and negative data are unlabeled.

Case-Control Scenario

In the case-control scenario, illustrated in Figure 2.5 and popular in many medical and environmental applications, the labeled positive and unlabeled data are collected separately from $p(x)$ often at different times and using different methods. An example of this could be viral detection. Blood samples of persons with known viral infection make up the positive case set. The background set is composed of samples previously collected from the general population. Some in the general population may have had the virus and been asymptomatic so the control set cannot be labeled negative, but rather should be labeled unknown.

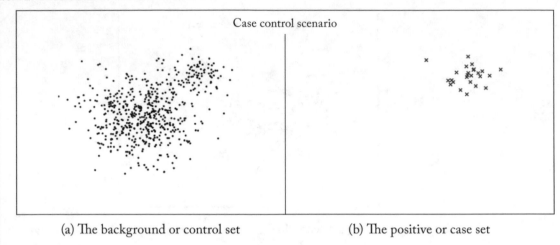

(a) The background or control set   (b) The positive or case set

Figure 2.5: The case-control scenario with data collected from two different datasets/sources.

The case-control scenario has two *potential* pitfalls for PU learning:

1. Duplicates between the positive case set and the control set must be eliminated or estimates of the class prior $\pi = p(y = 1)$ can be skewed. This may be difficult in some scenarios when unique identifiers are not available.

2. Depending on the time and difference in collection methods between acquiring the background control set and the positive case set, it is possible for a covariate shift to occur between the two sets. This is where the underlying data distributions change over time or due to events. This is most likely to be a problem if historical data is used for the background control set.

When the case-control scenario can eliminate duplicates and does not suffer from substantial covariate shift, the two sets can be combined and treated as a single training set. Otherwise, data from the two collection scenarios must be treated slightly differently.

2.3.2   DATA AND DATA LABELING ASSUMPTIONS

In addition to the data *collection* method, assumptions about the underlying structure and distribution of the data are required for most PU learning algorithms. Sometimes these assumptions are made implicitly, other times explicitly. In addition to assumptions on data distribution, assumptions on the selection of labeled positive data are required. How are the labeled positive data P_L selected or distributed regarding the set of all positive data P? The three most common assumptions, SCAR, SAR, and the Invariance of Order assumption, are described below after general data assumptions.

Data Distribution Assumption

As mentioned in Section 2.2.2, the underlying positive and negative data samples in P and N in our dataset are commonly assumed to be selected independently and identically from unknown underlying data distributions $p(x|y = 1)$ and $p(x|y = 0)$, respectively. While not always true, this is a simple, and typically implicit data assumption.

Data (Partial) Separability Assumption

Data separability refers to the separation of the class of interest from the remaining data in m-dimensional space. Data separability is dependent on the data's collected or engineered feature space. For example, data that is inherently three dimensional might be easily separable in three dimensions but not be separable if projected into only two dimensions. Well-separated data is shown in Figure 2.6a while inseparable data is shown in Figure 2.6b.

For fully labeled data, an effective supervised classification algorithm can determine data separability or partial separability as evidenced by its classification results. Feature engineering and deep learning can increase dimensionality by combining features to improve the likelihood of separability. When working with positive unlabeled data, this separability or partial separability must be assumed as it is impossible to determine if the data is well separated with homogeneous positive and negative classes with a low probability that a positive sample is labeled (Figure 2.6a) or if the data is poorly separated with substantial overlap and a high probability that a positive sample is labeled (Figure 2.6b). Both situations could result in the same positive and unlabeled set as illustrated in Figure 2.6c. Notice that in both cases, the final *PU dataset* has labeled and unlabeled data samples completely *overlapping and non-separable*. This is quite important and a fundamental reason as to why PU learning is an inherently difficult problem.

Partial separability, shown in Figure 2.7b, is sometimes referred to as the *positive subdomain assumption* [Bekker and Davis, 2020]. This is the assumption that the region of highest density of labeled positive samples must consist entirely of positive samples, as illustrated by the fully pink portion of Figure 2.7b. Where the density of known positive samples decreases, we conclude the overlap of positive and negative distributions. Either the partial or full separability assumption is typically made explicitly or implicitly in all PU learning algorithms.

Without expert domain knowledge, it not possible to differentiate between scenarios (a) and (b) when given the positive unlabeled set shown in (c) in Figure 2.6. Given the intractability of this problem, full or partial separability is explicitly or implicitly assumed in most every PU learning dataset.

SCAR: Selected Completely at Random Labeling Assumption

In addition to assumptions about the general data distributions and separability of the positive and negative datasets, all PU learning algorithms make some type of explicit or implicit assumption on the distribution of the labeled positive samples P_L as they relate to the set of all positive samples P.

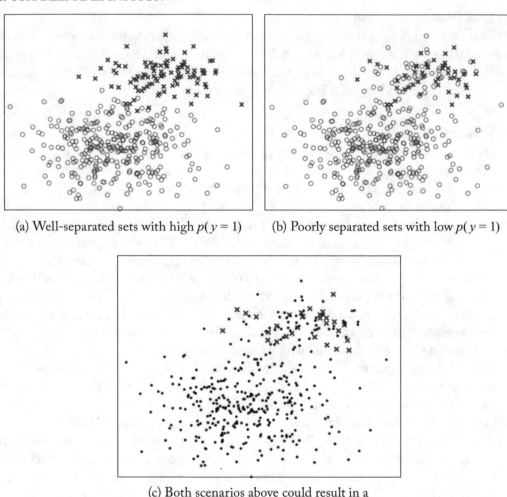

(a) Well-separated sets with high $p(y = 1)$ (b) Poorly separated sets with low $p(y = 1)$

(c) Both scenarios above could result in a
positive and unlabeled set like this one

Figure 2.6: Assumption of data separability or partial separability is necessary as positive unlabeled datasets with well separated positive and negative sets with a high positive class prior (a) are indistinguishable from poorly separated datasets with a low class prior (b). Both scenarios (a) and (b) could result in PU dataset (c).

The simplest (and most common) assumption is that there is no labeling bias, and the labeled positive samples have the same distribution as the set of all positive samples. Another way of saying this is that the labeled positive samples are selected completely at random from all positive samples. This is known as the *Selected Completely at Random* or SCAR assumption [Elkan

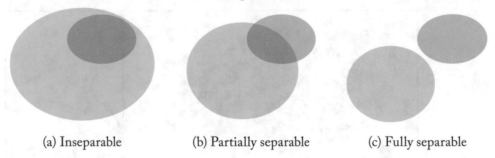

(a) Inseparable   (b) Partially separable   (c) Fully separable

Figure 2.7: Venn diagrams of inseparable, partially separable, and fully separable data.

Table 2.2: Properties of c under the SCAR assumption

$c = \dfrac{p(s=1\|x)}{p(y=1\|x)}$	The c variable equals the ratio of the conditional probability that a sample is labeled over the conditional probability that a sample is positive. This is proved in Section 4.1.5.
$c = \dfrac{p(s=1)}{p(y=1)}$	The c variable is closely related to the class prior. Significant work has gone into estimating either c or the class prior $p(y=1)$ directly from the dataset. This is explained in Section 4.4.
$c = \dfrac{\|P_L\|}{\|P\|} = \dfrac{\|P_L\|}{\|P_L\| + \|P_U\|}$	Under the SCAR assumption, the constant c estimates the ratio of labeled positive samples to all positive samples, which can also be written using set notation.

and Noto, 2008], and is illustrated in Figures 2.8 and 2.9b. The SCAR assumption can be stated mathematically using our label s variable from Section 2.2.2 as:

SCAR assumption

$$p(s = 1|\boldsymbol{x}, y = 1) = p(s = 1|y = 1) = c. \tag{2.6}$$

That is, c is the constant probability of a sample being labeled positive if it *is* positive. This is sometimes called the *labeling frequency*. Under this assumption, the set of labeled examples is selected independently and identically distributed from the set of all positive examples. While this assumption was first defined in Elkan and Noto [2008], it was previously used without definition in De Comité et al. [1999], Liu et al. [2003], Zhang and Lee [2005], and others.

The SCAR assumption ignores the problem of labeling bias which is not always realistic in practice. Any bias in the data is learned by the model, resulting in a biased model. Unfortunately, the nature of the PU learning problem often intensifies labeling bias. In real-world datasets, the "easier" positive samples furthest from the negative distribution are more likely to be identified and labeled. In medical disease diagnosis problems such as the diabetes example given at the

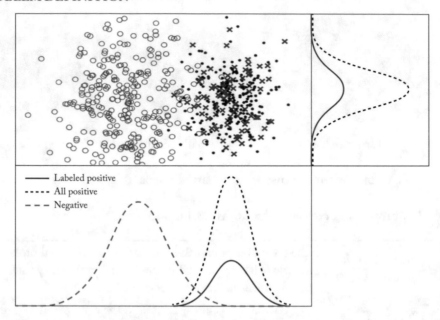

Figure 2.8: Illustration of the SCAR assumption in two dimensions. Notice that in both dimensions, the labeled positive samples are selected from the same distribution as all positive samples. Style credited to Bekker and Davis [2020].

beginning of the chapter, patients advanced in the disease and already diagnosed will be easiest to label. Borderline and early cases, the very cases we're often trying to identify using PU learning, will generally be more poorly represented. This means the labeled set is biased toward those with more advanced disease characteristics.

While SCAR assumes no bias, even datasets with bias can often be modeled successfully by algorithms making the SCAR assumption depending on the level and type of bias involved. In fact, most PU learning algorithms make use of the SCAR assumption either implicitly or explicitly.

While severe bias such as that in Figure 2.9d can be difficult or impossible to overcome entirely, recent work is being done to actively deal with and compensate for moderate levels of bias. Two such methods are described below.

SAR: Selected at Random Labeling Assumption

In the SAR assumption (*Selected At Random*), introduced in Bekker and Davis [2019], the authors weaken the labeling assumption by taking advantage of the propensity score, a concept developed by Rosenbaum and Rubin in 1983 [Rosenbaum and Rubin, 2006]. The propensity score is a balancing score that is used to reduce selection bias so that, conditional on the propensity score, the distribution on covariates is comparable between classes [Austin, 2011, Imbens

(a) True labels   (b) No bias

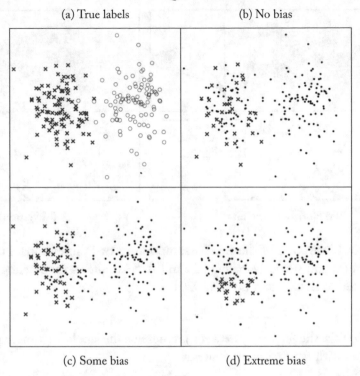

(c) Some bias   (d) Extreme bias

Figure 2.9: Examples of labeling bias. The ideal scenario (SCAR) with no bias is illustrated in (b), while a moderate amount of labeling bias is shown in (c). Extreme bias shown in (d) can make it impossible to learn a useful model.

and Rubin, 2015]. In traditional causal inference, the propensity score can be used to perform inverse probability weighting, which creates a pseudo population of weighted synthetic samples such that the distribution of overall measured covariates becomes class independent. This is a similar concept to using survey sampling weights to weight samples to ensure equal representation of the included populations. For example, if performing a telephone survey of 1000 people and only 375 of those surveyed were women, those 375 samples would be weighted more heavily in the survey results than the 625 men as women and men have approximately equal representation in the general population.

The SAR assumption uses this idea of a propensity score for class balancing but modifies it for the PU learning scenario in such a way as to allow for some labeling bias in some features as shown in Figure 2.10. This creates a more reasonable, if complex, assumption for real-world data collection and labeling situations. In this scenario, the labeling frequency is no longer assumed to be a constant c but is instead assumed to be a function $e(x)$ of its attributes or features.

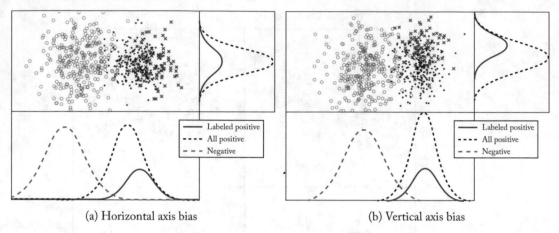

(a) Horizontal axis bias   (b) Vertical axis bias

Figure 2.10: Illustration of the SAR assumption in two dimensions. Horizontal labeling bias is shown in (a) and vertical labeling bias in (b). SAR allows for labeling bias in some features. Style credited to Bekker and Davis [2020].

To do this, the *propensity score e(x)* represents the label assignment probability for positive data samples with features or attributes x:

Propensity score

$$e(x) = p(s = 1|x, y = 1). \tag{2.7}$$

Under the SCAR assumption, this is a constant, but here it is allowed to vary based on the feature set of the data sample. This propensity score can be understood as an instance-specific probability that a data sample was selected to be labeled by the unknown and underlying labeling mechanism [Bekker and Davis, 2019].

The propensity score $e(x)$ in SAR differs from that in causal inference in that it applies only to positive samples—the probability of labeling a negative sample is zero in the PU problem. In some cases, the propensity score may be known due to the labeling mechanism being fixed and formalized through, for example, a hospital's protocol for testing people. This is more common in the medical field than in others.

For many datasets however, the propensity score is not known and must be estimated. Without any further assumptions, the problem is ill-defined so in order to estimate $e(x)$, the authors of Bekker and Davis [2019] make the assumption is made that it only requires a subset of the attributes $x_e \in x$, which they argue is often reasonable:

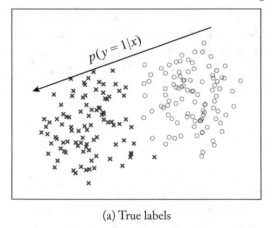

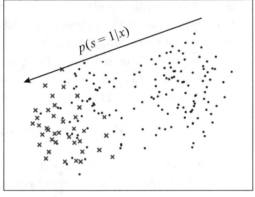

(a) True labels

(b) Selection bias under the invariance of order assumption

Figure 2.11: Illustration of the invariance of order assumption. The probability of a sample being labeled is NOT constant but instead increases with the probability that the sample is positive.

SAR assumption

$$p(s = 1|y = 1, \boldsymbol{x}) = p(s = 1|y = 1, \boldsymbol{x_e})$$
$$e(\boldsymbol{x}) = e(\boldsymbol{x_e}). \qquad (2.8)$$

Details of how this estimate is performed are described in depth along with the PU algorithm details from Bekker and Davis [2019] in Chapter 4.

Labeling Selection Assumption: Invariance of Order
In between the SCAR and the SAR assumptions is a less common assumption described in Kato, Teshima, and Honda [2019] and used in He et al. [2018] called the Invariance of Order assumption. The idea behind this assumption, also called order invariance, is that the sorted order over $\boldsymbol{x}$ created by the conditional probability of *labeling* a data sample as positive is the same as the sorted order of the conditional probability of that sample *being* positive. That is: *For any $\boldsymbol{x_i}, \boldsymbol{x_j} \in X$, we have*

Invariance of order assumption
$$p(y = 1|\boldsymbol{x_i}) \leq p(y = 1|\boldsymbol{x_j}) \iff p(s = 1|\boldsymbol{x_i}) \leq p(s = 1|\boldsymbol{x_j}). \qquad (2.9)$$

Another way of saying this is that the more likely a sample is to be positive, the more likely it is to be labeled positive. An example of this involving medical disease diagnosis was described when discussing bias in the SCAR assumption above.

This assumption is a little more difficult to work with than the other assumptions described above as, unlike the vast majority of other solutions in PU learning that use the single training set data acquisition model, Kato, Teshima, and Honda [2019] instead assumes the case-control data acquisition scenario (Section 2.3.1) and also assumes that outside domain knowledge is known about the dataset. That is, the class prior $p(y = 1)$ is assumed to be known, possibly allowing for control and case set overlap. These assumptions are dissimilar to most of the work in this field, the invariance of order assumption description is included here for completeness but will not be described in greater detail.

2.4   RELATED LEARNING PROBLEMS

In this section, we provide an overview of two problems that are so closely related to Positive Unlabeled learning that they are worth a small digression. First, *one-class classification* is described which includes *anomaly* and *novelty detection* along with many applications in *remote sensing*. Then *noisy classification*, which, in certain circumstances with class-specific noise, can be closely related or even reduced to PU learning. As these learning problems are substantial problems themselves, we are only able to briefly describe them here.

2.4.1   ONE-CLASS CLASSIFICATION

The PU learning problem is defined as a binary classification between two classes. One-class classification (OCC), sometimes called single-class classification (SCC), unary classification, or class modeling, is similar, but instead of distinguishing between two classes, a single class is identified from all others. OOC was first introduced by Moya and Hush [1996] with the intention of modeling a single class with no data from any other classes being available. This requirement is sometimes softened today, though unlike PU learning, many non-homogenous non-target classes are generally assumed to exist.

There are three common scenarios in OOC:

1. When data is available *only* for a single class as proposed in Moya and Hush [1996]. No other data, even unlabeled, is available belonging to any other class. This is most common in anomaly or novelty detection situations where correctly working or standard data is available and many potential unknown anomaly or novelty data could occur but does not yet exist [Jeon and Landgrebe, 1999, Khan and Madden, 2014, Schölkopf et al., 2000, Tax, 2002].

2. Some labeled data is available *both* for the class of interest and other, non-homogenous classes, though many data samples remain unlabeled and unknown. This frequently occurs when performing object detection in remote sensing applications [Deng et al., 2018, Guo et al., 2012, Khan and Madden, 2014, Liu et al., 2018, Muñoz-Marí et al., 2010].

3. Some data is available for the class of interest, and many unknown or unlabeled data samples from other, non-homogeneous classes, are available, though none are labeled. This situation is often grouped with number 2 above and can occur in remote sensing, health

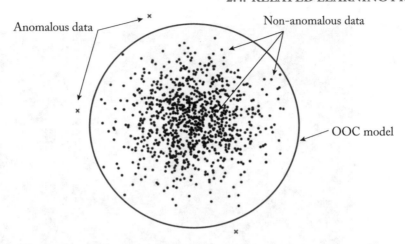

Figure 2.12: OOC model for anomaly detection. No anomalous data is available initially. A model of the data is created. Future anomalies or novelties do not fit within the model and are detected as not fitting.

applications, and others [Jeon and Landgrebe, 1999, Li, Guo, and Elkan, 2011a, 2020, Liu et al., 2018, Yu, 2005].

Scenario1 is the traditional and perhaps most common example of one-class classification. As mentioned, this often occurs in *Anomaly* and *fault detection*. In these problems, most of the data belongs to the non-anomalous class while only a small subset of the data, if any, is considered anomalous or abnormal. Many different problems can cause an anomaly in data or a fault in manufacturing and there is likely to be little to no anomalous data available before fault testing begins. In this situation, unlike most PU learning problems, non-anomalous data is the positive class of interest to be modeled—the anomalies and faults are typically caused by varied factors and are not homogenous. Because of this, binary classification and PU learning algorithms may not be as effective as other techniques specifically designed for this problem.

The traditional approach to the OOC problem is to ignore all data not belonging to the class of interest, so only the known positive data is modeled. Some common algorithms that are used for this purpose are One Class SVM (OCSVM), isolation forests, Gaussian mixture models for density estimation, and local outlier factors (LOF). Several sources including Chapelle, Schölkopf, and Zien [2006], Guo et al. [2012], Li, Guo, and Elkan [2011a], Liu et al. [2018], Scott and Blanchard [2009] have shown that using negative and unknown data can improve the results for this type of classification. When negative samples are available in addition to unknown samples, general semi-supervised approaches can be used [Muñoz-Marí et al., 2010], though Li, Liu, and Ng [2010] demonstrated that when heterogeneous, it may be better to ignore the negative training labels and treat all negative samples as unknown. When *only* unknown samples are unavailable to augment the set, PU learning is often the only option and can

Figure 2.13: An example of one-class classification for an autonomous vehicle. The green area marks the "clear to drive" area while all the objects in yellow boxes indicate obstructions. **Shutterstock: Scharfsinn.**

be quite effective. Recent work using PU learning for this problem is being done by Zhang et al. [2019].

Scenarios 2 and 3 in the list above can be thought of as a type of semi-supervised binary classification—the class of interest being labeled positive and all other classes being negative. When substantial labeled data from the negative classes is not available, this becomes a PU learning scenario. However, due to the heterogeneous nature of the negative "class," this is a more difficult learning scenario than binary PU with correspondingly lower confidence results.

Consider the problem of object identification for autonomous driving. An autonomous vehicle needs to identify many different types of objects—other vehicles, pedestrians, stop signs, animals in the road, and stop lights to name a few. To drive safely, the vehicle must classify video streams as being "clear to drive" or "obstruction." These labels indicate a binary classification, but the obstruction class consists of many different objects that are very dissimilar from one another, as shown in Figure 2.13. This task is more difficult than classifying something truly binary where each set is comparatively homogenous.

Remote sensing is a common OCC application [Li, Guo, and Elkan, 2020], and [Tsagkatakis et al., 2019]. In remote sensing applications, some type of land, water, or object is identified from images taken by satellite or some other high-flying aircraft. One application might be identifying water flooding areas after a large storm. Known lakes and rivers are labeled as water from standard maps (Figure 2.14a). All other areas would be unlabeled as

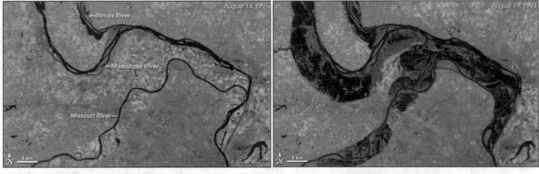

(a) Normal river conditions   (b) Flood conditions

Figure 2.14: A recording of the great flood of the Mississippi River, 1993 (St. Louis, Missouri). To identify floodwaters, known standard water sources shown in (a) would be labeled as water and all other areas as unknown for learning in (b). Because the unknown areas are non-homogenous, this is an OOC remote sensing problem, however PU learning is a common approach to solving these types of OOC problems. Landsat imagery courtesy of NASA Goddard Space Flight Center and U.S. Geological Survey.

they would normally be classified as land but may become flooded after a large rainfall. These might contain forests, deserts, grasslands, urban areas, and so on as shown in Figure 2.14. These areas of land are very different from one another, making this a one-class classification problem rather than a true binary classification. As we have some positive labels from the known water sources, and a large number of unlabeled regions that are usually dry but may become flooded (as shown in Figure 2.14b), this is also a great application of PU learning. Deng et al. [2018], Guo et al. [2012], Li, Guo, and Elkan [2011a], and Liu et al. [2018] are examples of papers using PU learning for remote sensing OOC applications.

The common approach to solving OOC problems using PU learning is to treat the labeled data from the class of interest as positive, and all other data as unknown. The negative class is the complement of the positive class, $N = P^c$. Standard PU algorithms can then be used, though the accuracy is expected to be somewhat lower than true two-class PU learning due to the non-homogenous nature of the negative (not from the class of interest) data. For more detail on using PU learning for OOC, Deng et al. [2018], Guo et al. [2012], Jeon and Landgrebe [1999], Li, Guo, and Elkan [2011a], Liu et al. [2018], and Yu [2005] are some great resources. A great general overview of OOC can be found in Khan and Madden [2014].

2.4.2   CLASSIFICATION WITH NOISY DATA

In most datasets, even fully labeled ones, data samples occasionally have incorrect feature values or are accidentally mis-labeled. These incorrect values are known as noise [Angluin and Laird, 1988]. There are two basic types of noise that are possible: attribute noise and class label noise.

Attribute noise is noise in the data sample's feature values x_i, where $i = 1, \ldots, n$, and can be due to human or machine error. For example, in a medical study, if a nurse misreads a thermometer when taking a patient's temperature, or if the thermometer itself was malfunctioning, the patient's recorded temperature would be incorrect. Attribute noise can also be caused by quantization error such as image pixilation or channel noise.

Class, or label noise, occurs when the recorded class label does not match the true class noise. In this case, the label $y^{true} \neq y^{recorded}$. For binary classification, this means $y^{true} = 1 - y^{recorded}$ as $y \in \{0, 1\}$. This kind of error can occur in several different situations, some of which are described below [Frénay and Verleysen, 2014, Kearns and Li, 1988, Natarajan et al., 2013].

1. Human error when labeling the data, caused by mistake or misunderstanding. As crowd-sourcing becomes more common, errors in data labels become more frequent due to in-consistency in the data collection process and lack of expert labelers.

2. Insufficient information. This can occur when existing test thresholds are not correct. For example, when Covid-19 testing began in 2020, the disease was just emerging, and several symptoms of the disease were not yet known. Therefore, testing criteria were incomplete, and patients were ill, but not diagnosed as such, or even tested due to the shortage of tests.

3. Subjective labeling can occur if class boundaries are not obvious, as when diagnosing things such as pain or stress levels. There is no measurement possible, any answers are entirely subjective and what might be very painful for one person might not be painful for another. Some illnesses such as fibromyalgia which are based on pain levels and other subjective criteria could be diagnosed or labeled differently by different physicians.

4. Malicious intent, which can occur in various situations such as food fraud [Barcaccia, Lucchin, and Cassandro, 2016] and fake reviews [Lappas, 2012], among others.

While both class and attribute noise can impact classifier accuracy, Zhu and Wu [2004] found that class noise is usually the more harmful of the two. This is likely because there are many features that can compensate for an incorrect one while there is only one class label, and the importance of each feature varies, while the label is always important [Frénay and Verleysen, 2014].

Class noise is often assumed to be evenly distributed. When the noise rate depends on the class value, the noise is said to be class conditional. If the noise is conditional entirely on the negative class as shown in Figure 2.15 such that all positive labeled data is positive and some labeled negative data is incorrectly labeled negative (the negative class is noisy), the problem can be immediately reduced to the PU learning problem. Anything labeled positive must be

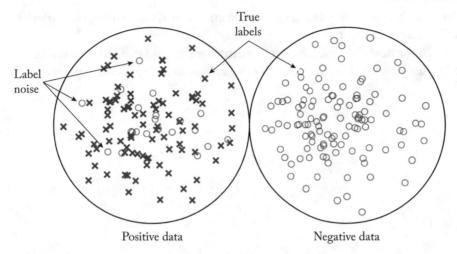

Figure 2.15: Example of class conditional noise. If all noisy labeled negative data (blue circles) were relabeled as unknown, this reduces to the PU learning problem.

positive while anything labeled negative has the potential to be misclassified and should instead be re-labeled as unknown.

Given a PU dataset with positive and unlabeled data, one PU learning strategy described in Section 4.1.3, is to assume that the unlabeled samples are noisy negatives with a relatively high probability of noise (being positive), while the positive samples are noiseless. This is the approach taken by Kearns and Li [1988], Lee and Liu [2003], Scott [2015], and Northcutt, Wu, and Chuang [2017]. Others, such as Liu et al. [2003], allow the user to specify the noise rates for both classes as a classification hyperparameter.

The problem of how best to deal with noisy data in the supervised learning environment is an active area of research. Coping with both noisy class labels [Frénay and Verleysen, 2014, Menon et al., 2015, Natarajan et al., 2013, Northcutt, Jiang, and Chuang, 2021] and attribute noise [Nettleton, Orriols-Puig, and Fornells, 2010, Zhu and Wu, 2004] are interesting fields of their own and there is substantial overlap between noisy classification and PU learning.

2.5   SUMMARY

In this chapter, we introduced the positive unlabeled (PU) learning problem and motivated its importance and prevalence in real-world classification problems. We provided a mathematical description of the problem and introduced the terminology and notation that will be used throughout this book. Further, we described two different data acquisition techniques, the single training set, and the case control scenario, along with five different assumptions that are sometimes made in this field. Finally, we ended the chapter with a discussion of two closely related problems—one-class classification and noisy classification, both of which sometimes use PU learning as a solution.

In the next chapter, Chapter 3, we discuss the unique evaluation challenges that exist when no labeled data exist from one of the two classes in a classification problem.

CHAPTER 3

Evaluating the Positive Unlabeled Learning Problem

Evaluating PU learning models poses challenges that are not present when evaluating standard supervised classification models. Because all negative data labels are missing in PU datasets, standard evaluation techniques that rely on calculating truth tables cannot be used as neither true negative samples nor false negative samples can be calculated. Because of this, neither a model's predicted precision nor accuracy can be calculated. Even the methods used to train PU models are different, as the standard supervised train—validate—test modeling technique is not possible when a substantial portion of the training dataset is unlabeled. Supervised classification uses Inductive learning to train a model that can be used on new, unlabeled data as shown in Figure 3.1a. In PU learning, as with many semi-supervised learning methods, either Inductive or Transductive learning is possible. The differences between these are summarized in Figures 3.1b and 3.1c.

Unlike supervised learning, PU learning classification *algorithms* and *models* are evaluated independently of one another. That is, identifying how an algorithm performs a task in relation to other algorithms is different than evaluating a specific algorithm's performance on a given dataset for hyperparameter tuning or model effectiveness. These two tasks are briefly described here and explained in more depth in the following sections.

- Task (1) Evaluating PU Algorithms.

  How does a PU *algorithm* compare with other PU algorithms? This comparison can include accuracy, computation complexity, runtime, memory required, and other metrics. This evaluation is done through algorithm profiling and benchmarking on *simulated* PU datasets as described in Section 3.1.

- Task (2) Evaluating PU Models.

  How well is a given PU *model* classifying a specific real PU dataset? In a real PU dataset, there are no negatively labeled data, so traditional metrics such as accuracy cannot be calculated and can only be estimated. PU model evaluation is performed differently than supervised learning and requires its own custom metrics and validation strategies. Details are given in Section 3.2.

The distinction between these two tasks is rarely discussed in the literature because supervised learning uses a single evaluation process and is far more common. However, PU learning

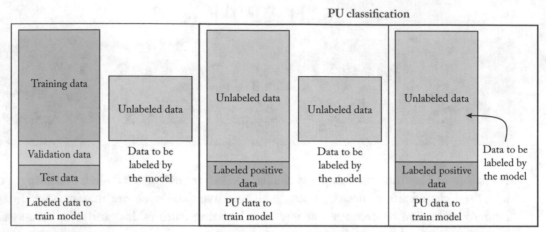

Figure 3.1: Data organization for (a) supervised learning and (b) and (c) PU learning datasets. Note that the training data includes unlabeled data in (b) and (c). Inductive learning creates a model from observed training data by learning general rules. These rules can then be applied to new data as shown in (a) and (b). Supervised learning is inductive. Transductive learning learns a model for the specific data on which it is trained and does not generalize to new data as shown in (c). Some PU learning models are exclusively transductive while others can create general models that can also be used inductively.

algorithms require two separate and distinct evaluation approaches because of the missing data in real PU datasets.

3.1   EVALUATING PU ALGORITHMS

The general way that ML algorithms are compared and evaluated against each other is through benchmarking. Benchmarking is the objective, unprejudiced comparison of multiple algorithms or methods over one or more commonly used datasets. The authors of Hoffmann et al. [2019] present a thorough breakdown of supervised classification benchmarking, detailing how supervised evaluation is often done in practice and formalizing a set of best practices. These include:

- using publicly available datasets rather than private, or difficult to access data;

- tests should be performed using procedures that are reproducible, unbiassed, and valid; and

- robust assessment requires consistent evaluation metrics over a broad spectrum and variety of datasets.

Unfortunately for accurate semi-supervised benchmarking (including PU learning), real datasets have missing data—meaning the performance can only be estimated (see Section 3.2). This means that true semi-supervised datasets, with natural, real-world bias, cannot be used effectively for comparing semi-supervised learning algorithms. Instead, to evaluate a semi-supervised learning algorithm and its properties, semi-supervised datasets must be simulated from supervised ones.

3.1.1   SIMULATING AND SYNTHESIZING PU DATASETS

A synthetic PU dataset can either be simulated from a fully labeled real-world dataset or created with synthetic data. There are advantages and disadvantages to both options. Using an existing, fully labeled dataset as a base allows for real-world data complexity. Biases can be added into the dataset on top of any existing data collection biases inherent to the data. The user has less control but more realism in this scenario. Synthesizing data allows greater control but generally less complexity and realism. Bias can be precisely added into the system if desired.

To simulate a PU dataset, a fully labeled supervised binary dataset is selected. Positive samples $x_p \in P$ are chosen to either be labeled and added to the P_L set or unlabeled and added to U using a labeling function $label\_pos(x_p)$. All negative samples $x_n \in N$ are added to U. True labels are stored for future evaluation. After performing PU classification, standard supervised evaluation metrics such as those briefly described in Section 3.1.2 can be used by recalling the true data labels.

The labeling function $label\_pos(x_p)$ can select positive samples to be labeled positive with a uniform random probability of $c \cdot |x_p|$ to satisfy the SCAR labeling assumption, with a probability of $e(x_p)$ to satisfy the SAR labeling assumption [Bekker and Davis, 2019], or with a probability proportional to a supervised classifier's probability of being labeled $\propto p(y = 1|x_p)$ for the Invariance of Order assumption [Kato, Teshima, and Honda, 2019]. These three labeling assumptions are described more fully in Section 2.3.2. Other biases can be introduced in non-standard ways as well.

Creating a synthetic dataset can be done by generating positive and negative data distributions $p(x|y = 1)$ and $p(x|y = 0)$ of the user's choice before applying the $label\_pos(x_p)$ function. These data distributions can be Gaussian [Kato, Teshima, and Honda, 2019], concentric circles [Hou et al., 2018], or spirals, moons, knots, or other shapes in between [Gong et al., 2019b, Ke et al., 2018, Kwon et al., 2020].

3.1.2   STANDARD ALGORITHM EVALUATION METRICS

In this section, we review standard one-dimensional classification evaluation metrics such as accuracy, error rate, f-score, precision, recall, and the Matthews correlation coefficient along with two-dimensional metrics such as the ROC and precision-recall (PR) curves. If you are already comfortable with these metrics, this section will be review and can be safely skipped.

		Actual class	
		Positive	Negative
Predicted class	Positive	True positive (TP)	False positive (FP)
	Negative	False negative (FN)	True negative (TN)

Figure 3.2: Confusion matrix for binary classification. The orientation is not fully standardized, and care must be taken to verify which cells are which in practice.

To compare PU learning *algorithms*, the algorithms of interest are run on simulated PU datasets. Because the true labels are known, the PU algorithm performance can be completely captured by standard supervised evaluation metrics such as those used for supervised classification, i.e., accuracy, error rate, and f-score, all of which can be calculated using the confusion matrix.

Confusion matrices, sometimes known as error matrices, are a special type of contingency table and a common method of visualizing and evaluating classification performance. In a confusion matrix (Figure 3.2), each column represents the true number of data samples belonging to the positive and negative sets, while the rows represent the predicted truth values.

Probably the most common supervised classification evaluation metrics are the accuracy and the error rate. Accuracy can be calculated from the confusion matrix as the number of elements correctly classified out of the total number classified.

Accuracy

$$\frac{\# \ classified \ correctly}{\# \ classified} = \frac{TP + TN}{TP + FP + TN + FN}. \tag{3.1}$$

The accuracy is a value between 0 and 1, typically represented as a percentage. The error rate is $1-$ Accuracy. If an algorithm correctly classifies data 98% of the time, its accuracy is 98% and its error rate is $1 - 0.98 = 2\%$.

Error rate

$$1 - Accuracy = \frac{FP + FN}{TP + FP + TN + FN}. \tag{3.2}$$

Unfortunately, accuracy and error rate are not effective when class sizes are unbalanced. If 99% of data samples are from class A and only 1% from class B, a classification algorithm that predicted that all samples belonged to class A would have an accuracy of 99% and be completely useless.

Positive unlabeled learning problems frequently have skewed and uneven class sizes. For example, an active area of PU research is working to identify genes in the human genome that may influence certain diseases. In this case, the class of interest—genes that influence the specified disease—will be far less than 1% of all genes in the genome.

Class balancing techniques such as oversampling of the smaller class, under sampling of the larger class, cost-sensitive learning, and data synthesis methods are commonly used to alleviate this class imbalance in supervised classification [Domingues et al., 2018, Elkan, 2001, Japkowicz and Stephen, 1998, Ling and Sheng, 2008]. However, as data labels are already reduced in PU learning, class balancing is generally not used to avoid biasing the limited labeled data. Because of this, metrics such as the f-score, the Matthews correlation coefficient, and the PR curve are recommended to replace accuracy and the ROC curve for PU learning.

The f-score (called the f1-score in statistics) is the harmonic mean of the precision and recall, which can be easily extracted from the confusion matrix. It is important to note that while these metrics are less sensitive to class imbalance than accuracy, they are not class symmetric and will return different values if the positive and negative classes are switched.

Precision calculates the probability that a data sample is positive given that it is *predicted* to be positive. Recall, also known as *sensitivity*, calculates the probability that a data sample is predicted positive if it *is* positive. The f-score is the harmonic mean of the precision and recall, balancing one against the other—being high when both are high and low if either is low.

PRECISION, RECALL, AND F-SCORE
Precision

$$p(y = 1|\hat{y} = 1) = \frac{TP}{TP + FP} \qquad (3.3)$$

Recall

$$p(\hat{y} = 1|y = 1) = \frac{TP}{TP + FN} = \frac{TP}{P} \qquad (3.4)$$

f-score

$$2 \cdot \frac{precision \cdot recall}{precision + recall} = \frac{2TP}{2TP + FP + FN}. \qquad (3.5)$$

Depending on the classification purpose, one may wish to prioritize precision over recall or vice versa. For example, when detecting a dangerous, yet rare disease, it may be preferable to err on the side of caution and prefer false positives over false negatives, improving the recall (sensitivity), over the precision.

A less common scalar classification metric is the Matthews correlation coefficient (MCC), also known as the *phi coefficient*. Though it was first introduced in 1975 [Matthews, 1975] (independently of the older phi coefficient described in 1912 [Yule, 1912]), this metric is relatively unknown by many ML practitioners, despite being simple and theoretically elegant. This metric computes the correlation coefficient between the true class and the predicted class after classification. The higher the correlation between the true class and the predicted class, the better the classification.

Matthews correlation coefficient (MCC)

$$\frac{TP \cdot TN - FP \cdot FN}{\sqrt{((TP + FP)(TP + FN)(TN + FP)(TN + FN))}}. \tag{3.6}$$

Unlike both the accuracy and f-score metrics, the MCC results in a number between -1 and 1 indicating the correlation between true and predicted classes. One advantage that the MCC has over the precision, recall, and f-score is that it is always symmetric, treating both the positive and negative classes as equally important. The MCC is unaffected by poorly balanced classes, creating a useful classification metric in all situations, balanced class sizes or not [Ramola, Jain, and Radivojac, 2019].

Until now we have been discussing "one-dimensional" performance metrics—metrics that summarize the performance of a model in a single numerical value. These are quite useful for quick comparisons and are extremely common when performing general algorithmic comparisons. Two-dimensional performance measures such as ROC curves (receiver operating characteristic curves) and PR curves (precision-recall curves) are often used to illustrate more nuance in performance than a single number can provide. ROC curves measure how a model's true positive rate varies as a function of its false positive rate as the decision boundary threshold changes. The PR curves measure a model's precision as a function of its recall similarly. These two-dimensional measures can be very effective in comparing PU learning algorithms, PR curves especially as they best handle skewed class sizes [Saito and Rehmsmeier, 2015].

Looking at the example ROC and PR curves in Figure 3.3, the best classifier in the ROC curve reaches into the far upper *left*-hand corner, and to the far upper *right*-hand corner for the PR curve. These two curves are consistent with one another if showing the results of five different classifiers on unbalanced datasets. ROC curves do not account for class imbalance, making the red classifier look ideal in Figure 3.3a. The PR curve in Figure 3.3b demonstrates that the green classifier is superior given that the dataset is unbalanced.

3.2   EVALUATING PU MODELS

In a real-world scenario with a real (non-simulated) PU dataset with actual missing labels, evaluating a PU model's performance becomes much more difficult. With PU data, the confusion matrix cannot be calculated but only estimated.

There are two goals involved with this task of PU model evaluation:

Goal 1: To create the best model, many algorithms require some form of hyperparameter tuning such as the learning rate, epochs, numbers of starting clusters, etc. This

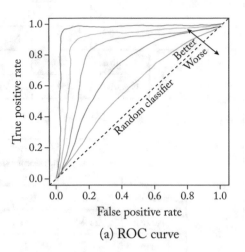

(a) ROC curve

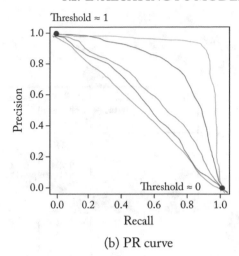

(b) PR curve

Figure 3.3: Classifier performance illustrated by a ROC curve (a) and a PR curve (b) on an unbalanced dataset. The colors of the classifiers are consistent in both curves demonstrating that in an unbalanced dataset, the PR curve will capture model performance better than a ROC curve.

requires that two models with slightly different parameters be compared for *relative* performance.

Goal 2: Ideally, having an estimate of ***absolute*** performance is useful for determining the confidence that should be placed in the model.

Determining the relative and absolute PU model performance are two very different tasks and very little appears to have been written on this subject in the literature, despite its importance. In this section we will present the most common evaluation techniques for evaluating PU models and describe which type, relative or absolute, of result the technique is useful for.

3.2.1 NAÏVE PU EVALUATION

In this book, we use the term "naïve" to mean ignoring the nuance of having unlabeled data samples and instead to treat all unlabeled samples as being negative. That is, it is naïve, though not necessarily uncommon, to treat s as a proxy for y. The advantage of this method is that it can be used to construct a confusion matrix from a validation set during training, as shown in Figure 3.4b. The confusion matrix will be biased toward underestimating performance (Figure 3.4c), but depending on the dataset and the labeling mechanism, this may be sufficient for *relative* model evaluation.

		Labeled class (s)	
		Positive	Negative
Predicted class ($\hat{y}$)	Positive	2	2
	Negative	0	4

(b)

True class label y	Label status s	Predicted probability $p(y=1)$	Predicted label $\hat{y}$
1	1	0.89	1
1	1	0.98	1
1	0	0.83	1
1	0	0.92	1
0	0	0.02	0
0	0	0.29	0
0	0	0.16	0
0	0	0.31	0

(a)

	Estimated performance (s)	True performance (y)
Accuracy	75%	100%
Error rate	25%	0%
F-score	0.667	1
MCC	0.577	1

(c)

Figure 3.4: Biased estimated performance metrics on an example PU dataset. In (a), a binary dataset with true class labels y and PU labels s is provided. True positive samples are unlabeled on the two shaded rows. The classifier correctly predicts every true label y so that $\hat{y} = y$. However, as y is unknown, the truth table in (b) uses s instead which results in the biased performance estimates shown in (c).

3.2.2   PU ESTIMATED F-SCORE (PUF-SCORE)

Probably the most commonly cited method in the literature is the one created in 2003 by Lee and Liu who proposed a performance criteria to evaluate *relative* model performance on real positive unlabeled datasets. As described in Section 3.1.2, an important property of the common f-score metric is that it is high when both the precision and recall are high, and low when either the precision or recall are low. The proposed performance criteria, which we will call a *Positive Unlabeled estimated f-score*, or "PUF-score," has similar properties and can be estimated from the validation set during training. Unlike the standard f-score which has a maximum value of 1, the *PUF*-score has no maximum value and thus is only useful for relative comparisons over the same data as with hyperparameter tuning. The derivation for this is as follows.
Notice that:

$$p(y = 1|\hat{y} = 1) \cdot p(\hat{y} = 1) = p(\hat{y} = 1|y = 1) \cdot p(y = 1), \tag{3.7}$$

so

$$\frac{p(y = 1|\hat{y} = 1)}{p(y = 1)} = \frac{p(\hat{y} = 1|y = 1)}{p(\hat{y} = 1)} \tag{3.8}$$

which can be re-written as

$$\frac{precision}{p(y = 1)} = \frac{recall}{p(\hat{y} = 1)}. \tag{3.9}$$

By multiplying each side by the recall, we get

$$\frac{recall \cdot precision}{p(y = 1)} = \frac{recall^2}{p(\hat{y} = 1)}. \tag{3.10}$$

Recall that the *f-score* is defined as

$$f\text{-}score = 2 \cdot \frac{precision \cdot recall}{precision + recall}. \tag{3.11}$$

Using Equations (3.7)–(3.10), Lee and Liu [2003] defined a new metric, which we call the PUF-score, to function similarly as:

$$PUF\text{-}score = \frac{precision \cdot recall}{p(y = 1)}. \tag{3.12}$$

This is sometimes abbreviated as $pr/p(y = 1)$ in the literature. Notice that $p(y = 1)$ is the class prior while $p(\hat{y} = 1)$ is an estimate of the class prior calculated as the percentage of labeled positive samples in the set X. Using the derivation shown in Equation (3.10), this simplifies to

$$PUF\text{-}score = \frac{recall^2}{p(\hat{y} = 1)}. \tag{3.13}$$

In Equation (3.4), we calculated the recall using just the positive set (recall = $TP/(TP + FP)$). While the entire positive set is not known in a real PU learning dataset, we can estimate the recall using the set of known labeled positive samples, P_L and the number of those that were correctly predicted by whatever PU learning algorithm was used, called TP_L. From these, we can estimate the recall as:

$$recall \approx \frac{TP_L}{P_L} = \frac{\sum_{s=1} \hat{y}}{\sum_{s=1} s}. \tag{3.14}$$

For hyperparameter tuning on a real-world dataset, a validation set rather than the full set of known positives should be used to minimize model bias.

3.2.3   ESTIMATION USING EXPECTATION

A paper by Jain, White, and Radivojac [2017] provides methods to estimate the performance of PU learning algorithms using the expectation of those algorithms over the long term. They point out that for a classifier $f : X \rightarrow \{0, 1\}$, the true positive rate (TPR), also called the recall, and the false positive rate (FPR) can be calculated using expectations with respect to the positive and negative distributions, as shown in Equations (3.15) and (3.16).

TRUE AND FALSE POSITIVE RATES

TPR/Recall

$$\mathbb{E}_P[f(x)] \tag{3.15}$$

FPR

$$\mathbb{E}_N[f(x)]. \tag{3.16}$$

However, given real finite datasets, without knowledge of these underlying data distributions, estimates of these values are required. The estimate of the TPR, or Recall, can be obtained as the empirical mean of the classifier f over the known positively labeled samples P_L as shown in Equation (3.17).

TPR/Recall estimate

$$\frac{1}{|P_L|} \sum_{x \in P_L} f(x). \tag{3.17}$$

Estimating the FPR and other metrics that rely on knowledge of true negative samples such as the precision, is not so simple. Jain, White, and Radivojac [2017] provide estimates of the false positive rate (FPR), the precision, and the AUC, as shown in Equations (3.18) and (3.19), where U represents the set of unlabeled samples in X containing both positive unlabeled (P_{UL}) and negative unlabeled, usually referred to as just negative samples (N).

ESTIMATES OF THE FPR, PRECISION, AND AUC FROM PU DATA

The positive class prior $\pi = p(y = 1)$, the TPR is calculated in Equation (3.17), and the AUC^{PU} is the area under the curve in an ROC curve treating the PU data as though it were positive/negative.

FPR$_{PU}$ estimate

$$\frac{1}{|U|} \sum_{x \in U} f(x) \tag{3.18}$$

FPR estimate

$$\frac{FPR_{PU} - \pi \cdot TPR}{1 - \pi} \tag{3.19}$$

Precision estimate

$$\frac{\pi \cdot TPR}{FPR_{PU}} \tag{3.20}$$

AUC estimate

$$\frac{AUC^{PU} - \frac{\pi}{2}}{1 - \pi}. \tag{3.21}$$

3.2.4   USING STANDARD EVALUATION METRICS

By making additional assumptions or knowing extra information about the dataset ahead of time, such as the class prior $p(y = 1)$ Claesen et al. [2015b], Jain, White, and Radivojac [2017], and Ramola, Jain, and Radivojac [2019] argue that standard evaluation metrics can be used for *relative* or *absolute* model evaluation with certain modifications. By using the class prior to compensate for the inherent performance bias illustrated in Figure 3.4, these methods attempt to normalize the resulting metrics which allows a degree of confidence in the final model.

The general idea from Claesen et al. [2015b] works as follows.

1. By making the SCAR assumption, we can estimate the class prior $p(y = 1)$.

2. If $p(y = 1)$ is known, then we can estimate the size of P as $p(y = 1) \cdot |PU\,dataset|$.

3. Given the size of P and the number of known positives (the size of P_L), we can estimate the number of latent positives in the unlabeled set, P_{UL}, as $|P_{UL}| = |P| - |P_L|$.

4. Because of the SCAR assumption, we can assume that the rank distributions of the labeled and unlabeled positives should be similar.

5. From both (3) and (4) above, we can calculate the total number of positive samples below or above a given rank, which is the necessary information needed to construct a confusion matrix [Bekker and Davis, 2020], from which the standard evaluation metrics discussed in Section 3.1.2 can be calculated.

Of course, we can simplify the entire problem and naively treat all unlabeled samples as negatives, which would allow us to construct a confusion matrix and use standard evaluation metrics directly. In Jain, White, and Radivojac [2017], the bias of this naïve approach is discussed. The authors demonstrate how this bias can be compensated for given estimates of the class priors. In Ramola, Jain, and Radivojac [2019], the bias of this naïve approach is described further and unbiassed estimates of the accuracy, balanced accuracy, f-score, and Matthews correlation coefficient are derived and provided. This dependent on knowledge or estimates of the class priors.

3.3   SUMMARY

To understand and compare PU learning algorithms, we first must know how to evaluate them. In this chapter we described why evaluating PU learning models is fundamentally different than evaluating standard supervised learning models and how there's a need for both PU algorithm and PU model evaluation. We reviewed some the standard evaluation metrics that are useful for evaluating PU algorithms on simulated and synthetic PU datasets where true labels are known for evaluation purposes. We then discussed the problems of relative and absolute model evaluation on true PU datasets and described the few methods from the literature. This is an underdeveloped area that could benefit from further study in the future.

CHAPTER 4

Solving the PU Learning Problem

The earliest papers on positive unlabeled learning were written in the late 1990s, such as Denis [1998] and De Comité et al. [1999]. Since that time, well over 150 papers, have been published, many of which are included in the bibliography of this book. In this chapter we survey the general types of algorithms that have been used for PU learning and explore some of the more common or important solutions in further detail.

> This section is intended as a general survey of the history of the PU field and its solutions. Algorithms will not have sufficient information included for reconstruction. References are provided if more detail is desired.

> Some selected algorithms include in-depth mathematical content that requires a deeper knowledge of advanced machine learning theory and principals than is expected of the general reader. This content is marked as advanced using the ML brain icon shown on the right and can be skipped without losing a high-level understanding of the field, its history, and its solutions.

When presented with a binary classification problem that uses labeled data samples belonging only to one class, the original tactic was to treat it as a one-class classification problem (see Section 2.4.1). Any unlabeled data was ignored and a model would be created exclusively from the labeled positive samples [Joachims, 1997]. New data was identified as either belonging to this model and labeled positive or being different from this model and labeled negative.

In 1998, Denis, and later Scott and Blanchard [2009], demonstrated that these models could be improved by using unlabeled samples (later, in 2016, Niu et al. provide theoretical evidence for PU learning to *outperform* traditional supervised learning under certain conditions). These arguments essentially initiated the semi-supervised subfield of PU learning in the late 1990s [De Comité et al., 1999, Denis, 1998, Nigam et al., 2000].

While many different types of PU learning algorithms exist, we organize them into the following solution categories.

Treating Unlabeled Data as Proxy Negatives
The generally more modern class of solutions treat the unlabeled samples as some sort of a proxy for negative samples. This can be done in many ways—naively or in a nuanced fashion through

weighting, *cost-sensitive learning*, *complex loss functions*, and so on. These are described in some depth in Section 4.1.

Two-Step Solutions

Another type of solution is called the *two-step* algorithm. The first step attempts to identify negative samples from the unlabeled set; the second step then uses an iterative semi-supervised learning algorithm on the known positive and probable negative samples. The two steps in this approach are generally independent of one another, allowing different algorithm combinations. Many older solutions fall in this category, though two-step solutions continue to be developed today. These are described in Section 4.2.

Other Methods and Considerations

Some recent PU learning methods do not fall easily into either of the above categories. They focus on issues of scalability and computational complexity or use reinforcement learning algorithms such as Generative Adversarial Networks (GANs) to synthesize new data to better learn data distributions. Ensemble methods will also be described in this section. These are described in Section 4.3.

At the end of this chapter, we address an important sub-problem in PU learning—that of estimating the positive class prior $\pi = p(y = 1)$. Many of the solutions described earlier in the chapter depend on an accurate estimate of this class prior which is often expected as an input to the algorithm. Only in recent years have algorithms been developed to effectively estimate the class prior from PU data.

Due to the limited amount of space available and the large number of papers published on this topic, we are forced to limit our selections and explanations below.

4.1   UNLABELED AS PROXY NEGATIVES

This section will include several different sub-categories of PU learning that use unlabeled samples as proxies for negative samples. As these unlabeled samples are often weighted in some way, algorithms of this type are sometimes referred to as belonging to a weighted approach. This is also sometimes called *biased learning* as certain assumptions, or biases (see Section 2.3), are made. Many of these algorithms can be defined as *cost-sensitive learning* because the cost of mis-labeling a known labeled positive sample is considered much worse than "mis-labeling" an unlabeled (proxy negative) sample as positive.

4.1.1   THE NAÏVE APPROACH

The most straightforward and naïve approach to modeling PU data is to treat all unlabeled samples as if they were negative and use standard supervised learning techniques to learn a classification model. The variable $s \in \{0, 1\}$ defined in Section 2.2 can be used to formally define this approach. Recall that $s = 1$ if a data sample is labeled positive and $s = 0$ if the sample is unlabeled. The naïve approach can then be formally stated as assuming that:

$$f(x) = p(y = 1|x) \approx p(s = 1|x). \tag{4.1}$$

The standard supervised learning algorithm can be a traditional SVM, logistic regression, a decision tree classifier, a neural network, or any other standard classifier. This is sometimes referred to as the "closed world assumption" [Bekker and Davis, 2020] in the field of Knowledge Representation. The only advantage of this technique is that it is simple, however we note that it is generally ineffective in practice. This approach is included for three reasons: (1) for completeness, (2) it is common in practice due to its simplicity, and (3) when combined with ensemble techniques (described in Section 4.3.2) it can sometimes become reasonably effective [Bekker and Davis, 2020, Mordelet and Vert, 2014].

4.1.2 NAÏVE BAYES AND OTHER BAYESIAN CLASSIFIERS

Despite having a similar name with the previous section, the Naïve Bayes (NB) algorithm's naivety is due to the "naïve" assumption of independence between feature values, rather than general model simplicity. Some early papers including Calvo, Larrañaga, and Lozano [2007] and Denis, Gilleron, and Tommasi [2002] have used various modifications on the standard Naïve Bayes algorithm to perform PU learning.

The standard NB classifier assigns each data sample x_i to the class $c = \{0, 1\}$ that maximizes its posterior probability $p(y = c_i|x_i)$ such that:

$$\hat{y} = \underset{c \in \{0, 1\}}{\operatorname{argmax}} \, p(c) \prod_{i=1}^{m} p(x_i|c). \tag{4.2}$$

To deal with PU data, Denis, Gilleron, and Tommasi [2002] modifies NB to create a *Positive Naïve Bayes* (PNB) algorithm that uses Laplace smoothing (also called additive smoothing) to estimate the likelihoods $p(x_i|0)$ and $p(x_i|1)$. The class prior $p(y = 1)$ is assumed known through domain knowledge to perform the calculation.

In Calvo, Larrañaga, and Lozano [2007], the PNB algorithm for PU learning is extended to improve the model using tree augmentation to construct a *Positive Tree Augmented Naïve Bayes* (PTAN) algorithm. An estimator for the class prior $p(y = 1)$ is provided using a Beta distribution to allow integration over all possible class priors. The authors in He et al. [2011] create PU versions of supervised Bayesian algorithms *Averaged One-Dependence Estimators* (AODE), *Hidden Naïve Bayes* (HNB), and *Full Bayesian network Classifier* (FBC) to create PAODE, PHNB, and PFBC, respectively.

4.1.3 NOISY NEGATIVES

We can improve upon the naïve approach described above by treating the unlabeled samples as *noisy* negatives, meaning that some positive samples have been mis-labeled. This reduces the

PU learning problem to that of noisy classification with class specific label noise [Frénay and Verleysen, 2014, Menon et al., 2015, Natarajan et al., 2013, Tanaka et al., 2018]. There are several noisy classifiers that can be used at this point, a selection of which are provided here at a very high level.

The PU learning problem is transformed into a noisy learning problem and solved by minimizing the weighted logistic regression error in Lee and Liu [2003], while the authors of Liu et al. [2003] offer a classifier they call the *Biased-SVM* with a modified cost function to account for the noisy data. More recent modifications of this from 2014 and 2018, respectively, include a *nonparallel support vector machine* (NPSVM) described in Zhang, Ju, and Tian [2014] and a *global and local learning* (GLLC) algorithm that uses a *biased least square support vector machine* (BLSSVM) in Ke et al. [2018]. Liu and Tao [2016] demonstrate that surrogate loss functions with importance reweighting allow any traditional classifier to be modified to work on noisy data. Gan, Zhang, and Song [2017] uses a tree augmented naïve Bayes algorithm called UPTAN (*Uncertain Positive Tree Augmented Naive Bayes*) to deal with the uncertainty of noisy data by creating a Bayesian network using dependence information among uncertain attributes to create a classifier. Scott [2015] uses *mixture proportion estimation* (MPE) to deal with noisy labels. The authors of Li et al. [2016] use a non-negative matrix factorization-based algorithm that incorporates known labels into an unsupervised structure via the consensus principle. The algorithm presented in Shao et al. [2015] assumes that most of the unlabeled data are negative and employs a Laplacian unit-hyperplane classifier to deal with the resultant noisy data. A more recent paper from 2018 converts the PU problem into a noisy risk minimization problem called *Loss Decomposition and Centroid Estimation* (LDCE) [Shi et al., 2018]. Another relatively recent paper, He et al. [2018], introduces the *probabilistic-gap PU model* (PGPU) that uses the invariance of order assumption described in Section 2.3.2 to allow for some labeling bias and treats all unlabeled data as noisy negative examples.

4.1.4 COST SENSITIVE LEARNING

Cost-sensitive learning allows misclassification error costs to differ by class. The PU learning problem can be thought of as a cost-sensitive learning problem where mis-labeling costs are highly skewed. In this method, all unlabeled samples are first labeled negative. The **cost** of mislabeling a true positive sample as negative (i.e., at least one of the unlabeled samples was positive but got labeled negative) is low, while the cost of mis-labeling a true negative sample as positive is high—the only samples labeled positive should **be** positive. These costs are typically represented in a cost matrix such as that shown in Figure 4.1. For PU learning, c_{01} would be relatively low while c_{10} would be very high and $c_{11} = c_{00} = 0$.

Elkan [2001] provides a foundation for cost-sensitive learning in which he shows that the threshold for an optimum cost-sensitive decision is p^* such that

$$\left(1 - p^*\right) c_{10} + p^* c_{11} = \left(1 - p^*\right) c_{00} + p^* c_{01}. \tag{4.3}$$

		Actual class	
		Positive	Negative
Predicted class	Positive	c_{11}	c_{10}
	Negative	c_{01}	c_{00}

Figure 4.1: Cost matrix C for a binary classifier. It should always be the case that $c_{10} > c_{00}$ and $c_{01} > c_{11}$. These are called the "reasonableness" conditions [Elkan, 2001].

When solved for p^*, this becomes

$$p^* = \frac{c_{10} - c_{00}}{c_{10} - c_{00} + c_{01} - c_{11}}. \tag{4.4}$$

Using this and assuming the "reasonableness" conditions described in Figure 4.1, an optimal binary prediction is

$$\hat{y}_{bin} = 1 \iff p(\hat{y}) \geq p^*. \tag{4.5}$$

From this, Elkan proves that to make a target probability threshold p^* correspond to a given probability threshold p_0, the number of negative samples in the dataset should be rebalanced by multiplying them by

$$\frac{p^*}{1 - p^*} \frac{1 - p_0}{p_0}. \tag{4.6}$$

In Elkan and Noto [2008], they applied this concept to the PU learning problem by treating each unlabeled sample as both a weighted positive a weighted negative sample at the same time.

> **Note Change in Notation—the rest of this section only**
> While this book generally uses the $y \in \{0, 1\}$ notation described in Sections 1.3 and 2.2 which is convenient in some problem formulations and is common in PU learning, when working with loss functions, the standard notation in the field instead uses $y \in \{-1, +1\}$. So, for the rest of this section only, positive data is represented by a label $y_{loss} = +1$ and negative data is represented by a label $y_{loss} = +1$. We chose to change the notation for this section only rather than modify the standard formulation to simplify further investigation into the literature. To differentiate this different notation, we will use y_{loss} and $\hat{y}_{loss}$ instead of y and $\hat{y}$ in this section.

In 2014, it was proposed in du Plessis, Niu, and Sugiyama [2014] that the PU learning problem can be solved using cost-sensitive learning between positive and negative data

Table 4.1: Cost-sensitive loss functions used for PU learning in the associated papers

Name	Definition	Convex	Used in
Ramp loss	$\ell(z) = \max\left\{0, \min\left\{1, \dfrac{1-z}{2}\right\}\right\}$	✗	(du Plessis, Niu, and Sugiyama, 2014)
Double hinge loss	$\ell(z) = \max\left\{0, \dfrac{1-z}{2}, -z\right\}$	✓	(du Plessis, Niu, and Sugiyama, 2015)
Sigmoid loss	$\ell(z) = \dfrac{1}{1 + \exp(z)}$	✓	(Kiryo et al., 2017)

using a non-convex ramp loss function. Notice that a loss function $\ell(\hat{y}_{loss}, y_{loss})$ where $\ell : \mathbb{R} \times \{-1, +1\} \to \mathbb{R}$ outputs the loss incurred by predicting an output $\hat{y}_{loss} = f(x)$ when the true label is y_{loss}. Table 4.1 provides the loss functions described in Kato et al. [2018], du Plessis, Niu, and Sugiyama [2014, 2015], and Kiryo et al. [2017].

The loss functions shown in Table 4.1 are unary. A unary loss function $\ell(z) = \ell(\hat{y}_{loss}, y_{loss})$ where $z = \hat{y}_{loss} \cdot y_{loss}$. These unary loss functions are standard in the literature. Given a loss function $\ell(z)$, the optimal classifier f can then be found by minimizing the classification risk, shown in Equation (4.7), defined as the expectation of the loss over the unknown joint density $p(x, y_{loss})$ [Kato et al., 2018].

Risk function

$$R(f) = \mathbb{E}[\ell(\hat{y}, y)]. \tag{4.7}$$

The authors of du Plessis, Niu, and Sugiyama [2014] argued that the non-convexity of the ramp loss avoided an intrinsic bias to the data. A year later, the same group was able to overcome the issue of concavity and create a non-biased convex formulation using the double hinge loss function [du Plessis, Niu, and Sugiyama, 2015] shown in the second row of Table 4.1.

With this loss function, du Plessis, Niu, and Sugiyama [2015] showed that the classification risk R_{PU} of an arbitrary decision function $f : \mathbb{R}^n \to \mathbb{R}^{nonneg}$ can be expressed as

$$R_{PU}(f) = \pi \mathbb{E}_p[\ell(\hat{y}_{loss}, +1)] - \pi \mathbb{E}_p[\ell(\hat{y}_{loss}, -1)] + \mathbb{E}_u[\ell(\hat{y}_{loss}, -1)], \tag{4.8}$$

where $\pi = p(y = 1)$ and $\mathbb{E}_p$ and $\mathbb{E}_u$ are the expectations over $p(x|y_{loss} = +1)$ and $p(x)$, respectively. Assuming no selection bias, i.e., when using the SCAR assumption, these expectations can be replaced with the corresponding sample averages.

A decision boundary function f can then be found by minimizing the classification risk $R_{PU}(f)$. Notice that knowledge of the class prior probability $\pi = p(y = 1)$ is assumed in Equation (4.8). Section 4.4 explores different methods of estimating the class prior.

Table 4.2: Classifier differences and definitions

Traditional classifier	$f(x) = p(y = 1 \mid x)$	The traditional classifier identifies the decision boundary between the *positive* and *negative* data points.
Non-traditional classifier	$g(x) = p(s = 1 \mid x)$	The non-traditional classifier attempts to identify the boundary between the *labeled* and *unlabeled* data points.

Further work published in 2017 in Kiryo et al. addresses some of the hyper-flexibility and overfitting problems from du Plessis, Niu, and Sugiyama [2015] by using a sigmoid loss (shown in Table 4.1) with a non-negative risk estimator to force a positive loss value.

Non-negative risk estimator

$$R_{PU}(f) = \pi \mathbb{E}_p[\ell(f(X), +1)]$$
$$+ \max\left\{0, \ \mathbb{E}_{X \sim p(x)}[\ell(f(X), -1)] - \pi \mathbb{E}_p[\ell(f(X), -1)]\right\}. \quad (4.9)$$

Using this non-negative risk estimator, the authors of Kiryo et al. [2017] were able to use a deep neural network to minimize the risk and construct a cost-sensitive PU model without having the same overfitting problems that had plagued du Plessis, Niu, and Sugiyama [2014] and du Plessis, Niu, and Sugiyama [2015].

4.1.5   USING A NON-TRADITIONAL CLASSIFIER

One of the foundational papers in PU learning Elkan and Noto [2008], introduced the idea of the non-traditional classifier, which continues to be developed today. As illustrated in Table 4.2, a traditional binary classifier defines the classification problem—a decision boundary or probabilistic classifier that effectively separates the positive from the negative data in the dataset. A non-traditional classifier attempts to construct a probabilistic boundary between the labeled and unlabeled data points in a PU learning problem. The non-traditional classifier by itself is computed the same way as the Naïve classifier described in Section 4.1.1, but unlike the Naïve classifier, calculating the non-traditional classifier in this section is only one step in a multi-step process.

Non-traditional classifiers have been shown [Elkan and Noto, 2008, Jaskie, Elkan, and Spanias, 2019] to be remarkably effective, relatively simple, and fast to implement and run, making them a good choice for PU learning. Their effectiveness depends on multiple factors, each of which is explained below.

The SCAR Connection:

Using the SCAR assumption (see Section 2.3.2) where the labeled positive data is selected completely at random from all positive data, Elkan and Noto [2008] proved that the desired PU traditional classifier is proportional to its non-traditional classifier divided by its labeling frequency, c. This proof is provided below.

Recall from Section 2.3.2 that the SCAR assumption can be formalized as:

$$p(s = 1|x, y = 1) = p(s = 1|y = 1) = c. \tag{4.10}$$

From this, Elkan and Noto [2008] proved that if the SCAR assumption were true,

$$f(x) = \frac{g(x)}{c} \tag{4.11}$$

which can also be written as

$$p(y = 1|x) = \frac{p(s = 1|x)}{c}. \tag{4.12}$$

The proof of this is given below.

Theorem 4.1   *Proof of Equation* (4.12) *by Elkan and Noto [2008].*
   Suppose the SCAR assumption holds. That is, the label s is independent of the feature values x such that $p(s = 1|x, y = 1) = p(s = 1|y = 1) = c$.
   Then

$$p(y = 1|x) = \frac{p(s = 1|x)}{c}.$$

Proof from Elkan and Noto [2008]: Consider $p(s = 1|x)$. As $y = 1$ whenever $s = 1$, this can be rewritten as:

$$
\begin{aligned}
p(s = 1|x) &= p(s = 1 \wedge y = 1|x) \\
&= p(y = 1|x) \cdot p(s = 1|y = 1, x) \\
&= p(y = 1|x) \cdot p(s = 1|y = 1) \\
&= p(y = 1|x) \cdot c.
\end{aligned}
$$

Divide both sides by c to complete the proof.

Equation (4.12) reduces the problem of correctly classifying the PU data to creating a naïve/non-traditional classifier and estimating c, the probability that a positive datapoint is labeled positive. There are two important consequences of this. As Elkan and Noto [2008] points out:

1. $f(x) = p(y = 1|x)$ is an increasing function of $g(x) = p(s = 1|x)$ which means that if the classifier is used only to rank data points by the likelihood of belonging to the positive class, the non-traditional classifier $g(x)$ can be used directly.

2. Additionally, $f(x) = g(x)/c$ is a well-defined probability, such that $f(x) \leq 1$ if and only if $g(x) \leq c$. Formally,

$$g(x) = p(s = 1|x) \leq p(s = 1|y = 1) = c. \tag{4.13}$$

This means is that $g(x)$ should never be greater than c.

The intuition for this second point was noted in Jaskie, Elkan, and Spanias [2019] as the labeled samples are selected completely at random from the positive samples, the labeled and non-labeled samples will be completely and totally non-separable from one another under the SCAR assumption. This means that at no point should the probability of a sample being labeled ever be 100%. If, for example, 30% of the positive samples are labeled ($c = 0.3$), then the highest probability that any sample should have of *being* labeled should be 30%.

Learning a Well-Calibrated Non-Traditional Classifier
Equation (4.13) states that the non-traditional classifier $g(x) = p(s = 1|x)$ should be bounded between 0 and c. However, well-calibrated classifiers such as logistic regression and probabilistic neural network classifiers [Metzen, 2015, Niculescu-Mizil and Caruana, 2005], which use the sigmoid output activation function shown in Equation (4.14), are designed to asymptote at 0 and 1 which makes them mis-specified for $g(x)$. Other classifiers such as SVMs and naïve Bayes must be calibrated before they can be used as probabilistic classifiers. Techniques such as Platt scaling [Platt, 1999] or isotonic regression [Zadrozny and Elkan, 2002] again produce output scaled between 0 and 1, not between 0 and c.

What this means is that when training a well-calibrated classifier to learn $g(x) = p(s = 1|x)$, if we were to plot the final probabilities of all the datapoints, the classifier attempts to "force" the probabilities to one. This can be seen in Figures 4.2a and 4.2b looking at the standard logistic regression algorithm (SLR) represented by the blue line where the true c values equal 0.25 and 0.75, respectively. In both cases, this standard classifier attempts to force an asymptote at 1, skewing the sigmoid shape past recognition. Notice that simply scaling the classifier between 0 and c would **not** produce nicely shaped sigmoid classifiers.

To solve this problem and obtain a non-traditional classifier that satisfies Equation (4.13), Jaskie, Elkan, and Spanias [2019] introduced what they called a *Modified Logistic Regression* (MLR) that is designed to asymptote at c rather than 1. This creates a "well-calibrated" *non-traditional* classifier that produces a well-defined final probability $f(x) = g(x)/c$. This MLR classifier performance is illustrated by the orange lines in Figure 4.2. While standard logistic regression and probabilistic neural networks use a sigmoid function shown in Equation (4.14), the MLR non-traditional classifier uses a modified sigmoid function with a variable upper bound

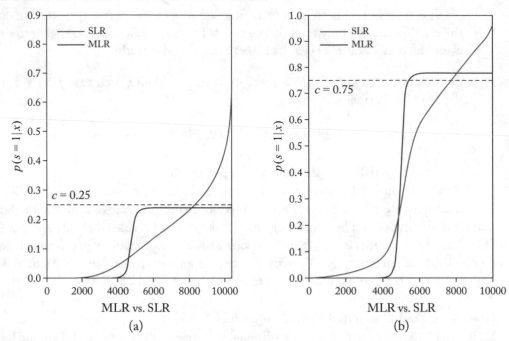

Figure 4.2: Output of two non-traditional classifiers $g(x) = p(s = 1|x)$ showing the probability that a datapoint is labeled on a simulated dataset. In (a) $c = 0.25$ and in (b) $c = 0.75$. Exactly half (5000) the data was positive driving the vertical component of the sigmoid. A standard classifier like logistic regression (blue curve) attempts to reach a probability of 1, while Jaskie, Elkan, and Spanias [2019]'s MLR algorithm (orange curve) asymptotes at or near c.

shown in Equation (4.15). This is a simplification of a Generalized Logistics Curve, also known as a Richard's Curve [Richards, 1959].

Sigmoid function

$$p(s = 1|x) = \frac{1}{1 + e^{-\overline{\omega} \cdot \overline{x}}} \qquad (4.14)$$

Modified sigmoid function for MLR

$$p(s = 1|x) = \frac{1}{1 + b^2 + e^{-\overline{\omega} \cdot \overline{x}}}. \qquad (4.15)$$

Jaskie, Elkan, and Spanias [2019] used a stochastic gradient ascent algorithm to perform a maximum likelihood estimation (MLE) to estimate the learned variables $\overline{\omega}$ and b in Equation (4.15):

$$MLE = \underset{b,\,\bar{\omega}}{\operatorname{argmax}} \sum_{i=1}^{m} \log \left(\left[\frac{1}{1 + b^2 + e^{-\bar{\omega}\cdot\bar{x}_i}} \right]^{s_i} \left[1 - \frac{1}{1 + b^2 + e^{-\bar{\omega}\cdot\bar{x}_i}} \right]^{1-s_i} \right). \qquad (4.16)$$

Estimating the Labeling Frequency c

As noted above, Equation (4.12) reduced the problem of creating a traditional classifier $f(x) = p(y = 1|x)$ to the problem of learning a non-traditional classifier $g(x) = p(s = 1|x)$ and estimating the labeling frequency c. While there are several methods of estimating c and the closely related class prior $p(y = 1)$ (see Sections 2.3.2 and 4.4, respectively), Jaskie, Elkan, and Spanias [2019] found that the modified sigmoid function in the MLR algorithm shown in Equation (4.15) estimates c directly from the dataset as the upper asymptote. This can be seen in Figure 4.2 and calculated as

$$\hat{c} = \frac{1}{1 + b^2}. \qquad (4.17)$$

As noted in Equation (4.13), the non-traditional classifier $g(x)$ should be less than or equal to c. As shown in Figure 4.2, the asymptote of $g(x)$ will not always be exactly at c, depending on the actual sampling of labeled samples. In the simulated dataset used in this example, the random sampling of the positive set in Figure 4.2a was well distributed while the random sampling of the positive set in Figure 4.2b randomly labeled more datapoints near the decision boundary than away from it, leading to a slightly elevated estimate of c. It is important to note that even when the SCAR assumption is enforced during simulation, the randomness of the selection process itself can induce some variability.

Pulling it All Together

With a calibrated non-traditional classifier $g(x) = p(s = 1|x)$ such as that shown in Equation (4.15) an estimate of c such as that calculated as part of developing $g(x)$ and shown in Equation (4.17), an effective PU classifier $f(x) = p(y = 1|x)$ can be created using Equation (4.12). This process can be seen in Figure 4.3.

4.1.6   DEALING WITH BIAS

One of the limitations of the non-traditional classifier described in the previous section is that it relies on the SCAR assumption and thus is unable to accommodate labeling bias. As we saw in Section 2.3.2, two alternative labeling assumptions have been proposed to cope with labeling bias—the *Selected at Random*, or SAR assumption, and the *Invariance of Order* assumption.

Using the SAR Assumption

Recently, two important papers Bekker and Davis [2018b, 2019] introduced the SAR (selected at random) assumption and its application for positive unlabeled data with a selection bias. The SAR assumption was described in Section 2.3.2, but essentially says that instead of the labeling

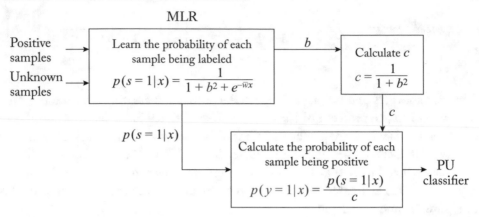

Figure 4.3: Putting it all together: the three steps a non-traditional classifier using the MLR sigmoid function.

Table 4.3: Differences between the SCAR and SAR assumptions

SCAR	$p(s = 1|x, y = 1) = p(s = 1|y = 1) = c$	Constant c
SAR	$p(s = 1|x, y = 1) = e(x)$	Propensity score $e(x)$

frequency being assumed to be constant as in SCAR, it is assumed to be a function of the data attributes/features called the *propensity score* $e(x)$.

In some situations, in the medical field particularly, the propensity score $e(x)$ is known and an unbiassed classifier can be trained by either scaling non-traditional output probabilities by $1/e(x)$ instead of $1/c$ or through integration in the training process. When the propensity score is known, a cost (or risk) function such as that shown in Equation (4.8) can be generalized to work with the propensity score rather than the class prior $\pi = p(y = 1)$ [Bekker and Davis, 2019].

More commonly, the propensity scores for the data are not known in advance and must be estimated. Unfortunately, as given, this problem is ill-defined. Without the SCAR assumption, any unlabeled sample could be explained by either a low positive class probability or a low labeling probability. To deal with this, Bekker and Davis [2018b, 2019] assume that the propensity function only requires a subset of the attributes $x_e \in x$:

$$p(s = 1|y = 1, x) = p(s = 1|y = 1, x_e)$$
$$e(x) = e(x_e).$$

(4.18)

Given this assumption and that of partial data separability described in Section 2.3.2, Bekker and Davis [2018b, 2019] propose an Expectation Maximization (EM) algorithm they call SAR-EM which is described briefly below.

Data is assumed to be generated by the following process where x_e are the propensity attributes:

$$
\begin{aligned}
(x, y, s) &\sim p(x, y, s) \\
&\sim p(x) \cdot p(y|x) \cdot p(s|x, y) \\
&\sim p(x) \cdot p(y|x) \cdot p(s|x_e, y).
\end{aligned}
\tag{4.19}
$$

They assume that the process of classifying and labeling the data samples x can be modeled using parameters θ and ϕ, respectively:

$$
(x, y, s) \sim p(x) \cdot p(y|x, \theta) \cdot p(s|x_e, y, \phi).
\tag{4.20}
$$

The goal of SAR-EM is to find θ and ϕ that best explain the observed variables x and s.

The expectation and maximization steps are typically iterated until convergence, or the maximum number of iterations is reached. Without going through the full derivation, we provide the expectation and maximization equations here:

Expectation:

$$
p(y = 1|x, s = 1, \theta, \phi) = 1
$$

$$
p(y = 1|x, s = 0, \theta, \phi) = \frac{p(y = 1|x, \theta)\, p(s = 0|y = 1, x_e, \phi)}{p(y = 1|x, \theta)\, p(s = 0|y = 1, x_e, \phi) + p(y = 0|x, \theta)}.
\tag{4.21}
$$

Maximization:

$$
\begin{aligned}
&\max_{\theta, \phi} \mathbb{E}_{y|x,s,\theta,\phi} \log p(x, s, y|\theta, \phi) \\
&= \max_{\theta} \mathbb{E}_{y|x,s,\theta,\phi} \log p(y|x, \theta) + \max_{\phi} \mathbb{E}_{y|x,s,\theta,\phi} \log p(s|y, x_e, \phi).
\end{aligned}
\tag{4.22}
$$

Using the Invariance of Order assumption

Using the *Invariance of Order* assumption defined in Section 2.3.2, Kato, Teshima, and Honda [2019] deal with selection bias by minimizing the classification risk function, assuming that the class prior $p(y = 1)$ is known. This approach is quite similar to the cost-sensitive solutions described in Section 4.1.4 and used in papers by du Plessis, Niu, and Sugiyama [2014, 2015b] and Kiryo et al. [2017]. However, the authors of Kato, Teshima, and Honda [2019] assume a selection bias and therefore cannot create an empirical classification risk function. Instead, they use a *partial identification* technique to extract some useful information from a function, without attempting to identify the entire function. To do this, they define a score function

$$
r(x) = \frac{p(x|y = 1, s = 1)}{p(x)},
\tag{4.23}
$$

and then introduce the following theorem on the order preserving property of the score function. The proof of this theorem is not provided here, but can be found in Kato, Teshima, and Honda [2019].

Theorem 4.2   *Order-preserving property of the score function in Equation* (4.23)

Suppose that the Invariance of Order assumption given in Equation (2.9) *and reproduced below holds.*

Invariance of Order Assumption

$$p(y = 1|x_i) \leq p(y = 1|x_j) \iff p(s = 1|x_i) \leq p(s = 1|x_j).$$

Then, for any $x_i, x_j \in X$,

$$p(y = 1|x_i) \leq p(y = 1|x_j) \iff r(x_i) \leq r(x_j).$$

To estimate Equation (4.23), Kato, Teshima, and Honda [2019] suggests either minimizing a *pseudo-classification risk* using the logarithmic loss function $\ell(\hat{y}, y) = -y \log(\hat{y}) - (1 - y) \log(1 - \hat{y})$ or performing *direct density ratio estimation*.

Using the non-negative loss function proposed by Kato, Teshima, and Honda [2019], Kiryo et al. [2017] minimizes the pseudo classification risk where $\mathbb{E}_p^{bias}$ is the expectation over $p(x|y = 1, s = 1)$. This can be estimated with $\widehat{\mathbb{E}}_p^{bias}$ being the empirical mean of the positive dataset and $\widehat{\mathbb{E}}_u$ being the empirical mean of the unlabeled dataset. In the following equations, $\mathcal{H}$ is a subset of a set of measurable functions, known as a hypothesis set.

Pseudo classification risk

$$\hat{r} = \frac{1}{p(y = 1)} \underset{f \in \mathcal{H}}{\text{argmin}} \left[-p(y = 1) \widehat{\mathbb{E}}_p^{bias}[\log f(X)] \right.$$

$$\left. + \max\left\{0, p(y = 1) \widehat{\mathbb{E}}_p^{bias}[\log(1 - f(X))] - \widehat{\mathbb{E}}_u[\log(1 - f(X))]\right\} \right]. \tag{4.24}$$

As an alternative, Kato, Teshima, and Honda [2019] proposes a direct empirical density ratio estimation for estimating $r(x)$ using what they call an *unconstrained Least-Squares Importance Fitting* (uLSIF) which has a closed form solution. To do this, a class T of non-negative measurable functions $t : \mathbb{R}^n \to \mathbb{R}^{nonneg}$ is introduced and the squared error between t and r is considered. As explained in Kato, Teshima, and Honda [2019], this can be used to obtain an empirical estimate for the score function r.

Density ratio estimation using uLSIF

$$\hat{r} = \operatorname*{argmin}_{t \in \mathcal{H}} \left[\frac{1}{2} \widehat{\mathbb{E}}_u[(s(X))^2] - \widehat{\mathbb{E}}_p^{bias}[s(X)] \right].$$

(4.25)

Using this estimate of the score function r and Theorem 4.2, Kato, Teshima, and Honda [2019] estimate a threshold value from the dataset and the class prior to rank and predict $\hat{y}$.

4.2   TWO-STEP SOLUTIONS

One relatively straightforward approach in dealing with PU data is to first identify probable negative data samples N_P from the unlabeled set. With these probable negative samples and the known samples, the problem is then reduced to a supervised or semi-supervised learning problem and can be solved using a variety of methods. This idea is called the two-step approach due to the two steps in the algorithm.

1. Identify probable or likely negative samples N_P.

2. Use standard supervised or semi-supervised classification algorithms on the probable negative N_P and known positive P_L samples.

These steps can be performed in either a single pass or in an iterative fashion. These two steps can also be considered independently, allowing for a variety of algorithmic combinations as described in detail in survey papers by Bekker and Davis [2020], Zhang and Zuo [2008], and Kaboutari, Bagherzadeh, and Kheradmand [2014]. While the two-step approach was more common in the early days of PU learning, work does continue to be done in this area such as that in He et al. [2018] and Zhang et al. [2019].

4.2.1   STEP 1: IDENTIFYING LIKELY NEGATIVES

Many heuristics have been used to identify probable negative samples from PU datasets, often making use of some form of distance or similarity calculation between the data samples. In this section, we will attempt to provide high-level descriptions of several of these methods—references are provided for further details. Many of the earlier algorithms were originally designed for text classification [Fung et al., 2006, Li and Liu, 2003, Li, Liu, and Ng, 2007, Liu et al., 2002, Liu and Peng, 2014, Peng, Zuo, and He, 2008, Yu, Han, and Chang, 2002, Yu, Man, and Chang, 2004, Yu and Li, 2007, Zhang and Zuo, 2009]. The types of heuristic algorithms for identifying probable negative samples will be broken into four general categories: probabilistic measures, text feature frequency, general distance and clustering metrics, and graphing techniques.

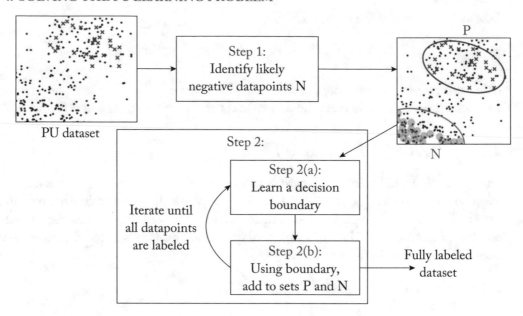

Figure 4.4: Two-step algorithm general structure/block diagram.

Probabilistic Measures

To identify reliable negative samples, a probabilistic algorithm can be used to calculate the chance that a data sample belongs to the positive class. One common algorithm used for this purpose is naïve Bayes (NB) (see Section 4.1.2).

- One of the earliest two-step algorithms, the S-EM (or *Spy Expectation Maximization*) was published by Liu et al. [2002] and described in further detail in Section 4.2.3. It uses this method along with known positive samples acting as spies to determine a threshold cutoff value for identifying likely negative samples.

- The PE_PUC (or *Positive document Enlarging PU Classifier*) was published by Yu and Li in 2007 and begins with a similar approach, extracting reliable negatives from the unlabeled set using NB. Unlike S-EM, however, PE_PUC doesn't stop there but continues by extracting reliable *positive* samples from the remaining unlabeled using a semi-supervised graph algorithm. This is described further in the graph section below.

- A more recent paper from 2018, PGPU (*Probabilistic-Gap PU Model*) from He et al. [2018] estimates the difference or *gap* between the Bayesian posteriors $p(y = 1|x)$ and $p(y = 0|x)$. The smaller the gap, the more difficult it is to label the sample, which is done using a Bayesian optimal relabeling method which produces an initial, noisy labeling of all the unlabeled data.

Text Feature Frequency

While general probable negative identification methods are relevant to any type of data, early PU learning methods were developed in a text classification framework and many early algorithms focused on text feature extraction. Text classification algorithms often use a bag-of-words representation with each word representing a feature. A natural text-specific metric looks at the frequency of the word, or features, in each document. This is done in one of several different ways. One of the most common representations is called 1-DNF, or a monotone disjunction list (a list of individual words with a logical OR between them).

- An early PU learning algorithm, PEBL (or *Positive Example Based Learning*) [Yu, Han, and Chang, 2002, Yu, Man, and Chang, 2004] uses 1-DNF to identify which features occur least often in the positive set to identify probable negative documents.

- Peng, Zuo, and He [2008], created an "improved" 1-DNF where a feature is identified as a "positive" feature if its frequency in the known positive set is greater than its frequency in the unlabeled set and its overall frequency is greater than some threshold, set through experimentation.

- PNLH [Fung et al., 2006], introduced *Positive Examples and Negative Examples Labeling Heuristics*, defines a similar though not identical concept as 1-DNF, to extract a set of reliable negatives using the differences between the feature distributions between the positive and unlabeled sets.

- In a different approach, the LGN algorithm (*Learning by Generating Negative examples*) published by Li, Liu, and Ng [2007] identifies which words are least likely to belong to a positive document using their conditional probability, and uses these to construct a single negative document. This provides a clear negative sample for classification.

General Distance and Clustering Metrics

The general idea behind selecting reliable negative samples is to select samples that are not easily confused with the positive set. A straightforward way to do this is to choose whichever unlabeled samples are farthest away from the ones that are known to be positive using some sort of a distance measure. This concept is illustrated in Figure 4.5.

- One of the earliest PU learning algorithms, Roc-SVM, published in 2003, used a Rocchio nearest centroidclassifier applied to text documents that are represented by TFIDF (*term frequency—inverse document frequency*) feature vector [Li and Liu, 2003]. Unlabeled documents that are closer to the unlabeled centroid are labeled as probable negative samples.

- Zhang and Zuo [2009] also focus on text documents and represent them by their TFIDF feature vectors. However unlike Roc-SVM, Zhang and Zuo [2009] calculate

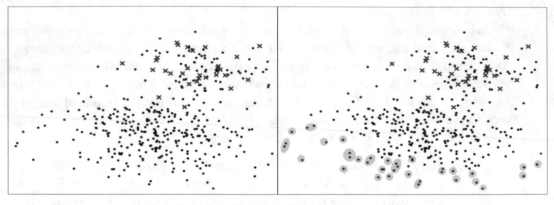

(a) PU dataset with labeled positives shown as red x's and the unlabeled as black dots

(b) PU dataset with reliable negatives chosen as those most distant to the positive set

Figure 4.5: Selection of reliable negative samples using simple distance metric such as kNN.

and rank documents by similarity by taking the cosine of the normalized document vectors. They then use a k-Nearest-Neighbor (kNN) algorithm on this ranked data to identify probable negative samples.

• Another algorithm that uses kNN is the more recent Pulse algorithm [Ienco and Pensa, 2016], *Positive Unlabeled Learning for Categorical datasEt*, focuses exclusively on categorical data as the name suggests. To select probable negative samples, a technique called DILCA or *Distance Learning for Categorical Attributes* from Ienco, Pensa, and Meo [2012] is used to compute a context-based distance between any pair of values of the target attribute based on the similarity between their probability distributions. The kNN algorithm is then used to select samples farthest from the known positives to be labeled as probable negatives.

• Performing unsupervised clustering of the PU dataset and then selecting data from clusters farthest from the positive data is a method that Chaudhari and Shevade [2012], Jeon and Landgrebe [1999], and Liu and Peng [2014] have used. One of the earliest algorithms was published by Jeon and Landgrebe [1999] who proposed a weighted clustering approach. A later MCLS, or *Maximum Margin Clustering with Least Squares SVM*, algorithm from 2012 was introduced by Chaudhari and Shevade [2012] that uses k-means to cluster the data. C-CRNE [Liu and Peng, 2014] (*Clustering-based method for Collecting Reliable Negative Examples*) used hierarchical agglomerative clustering to extract likely *positive* samples from the unlabeled set. The remaining unlabeled data were considered to be reliably negative. They did this on text documents using a modified TFIDF called TFIPNDF (*Term Frequency Inverse Positive-Negative*

Document Frequency), a procedure that weighted features differently depending on how frequently they occur in the positive and negative documents.

Graph Methods

Graph algorithms are a less common method for identifying probable or likely negative data in two-step algorithms, but some recent work in Gong et al. [2019b] looks promising.

- A text document classification algorithm named PE_PUC, or *Positive document Enlarging PU Classifier*, proposes an interesting two-part strategy for the first part of their two-step algorithm [Yu and Li, 2007]. First, they identify reliable negative documents from the unlabeled set using a naïve Bayes algorithm. However, unlike other two-step PU algorithms, they do not stop but next extract reliable positive documents from the remaining unlabeled samples using a semi-supervised graphing algorithm that computes and propagates node similarity from the known positives. This is especially useful if the set of known positive samples is very small.

- A recent algorithm, MMPU (*Multi-Manifold PU learning*), published in Gong et al. [2019b], uses graphs to represent the similarities in the feature space including local similarity, structural similarity, and semantic similarity between pairs of examples. These graphs are used to identify smooth, low-dimensional manifolds, each of which corresponds to a class. Likely negative samples are chosen from the manifold corresponding to the negative class. The purpose is to include all structural relationships between the unlabeled data rather than considering each unlabeled sample as independent from all others as is done in some other methods.

4.2.2   STEP 2: SUPERVISED CLASSIFICATION

While there are many different heuristics used to identify probable negative samples from the unlabeled set, in the PU literature, only a few semi-supervised learning algorithms are commonly used to create classifiers from the resulting probable negative samples N_P and known, labeled positive samples P_L. The most common is an iterative SVM or a variant thereof. Alternatively, the naïve Bayes (NB) algorithm is paired with the expectation maximization (EM) algorithm to accommodate missing labels. While it is possible to ignore the unlabeled data and use a fully supervised learning algorithm on the known positive samples and the reliable negatives selected in step 1, this ignores valuable information that could improve learning [Denis, 1998]. An exception to this is the PGPU algorithm in He et al. [2018], as it predicts noisy labels for all unlabeled sample during step 1, and uses a single iteration of a weighted SVM to create a final classifier.

The iterative SVM and NB with EM algorithms will be briefly described here, along with their original algorithmic applications. As the two steps in a two-step PU learning algorithm are generally independent, algorithm selection can be matched in different combinations than described in the literature.

Iterative SVM

The idea behind an iterative SVM is to take the labeled data available (both positive and probable negative—ignoring the unlabeled) and to classify it using a standard SVM (support vector machine). At this point, the remaining unlabeled samples are inspected. Unlabeled samples that fall beyond the margin and solidly and unambiguously into the negative class are added to the negative set. In the case of the MCLS algorithm [Chaudhari and Shevade, 2012], unambiguously positive samples are also added to the positive set. A new classification iteration is performed. This process is repeated until all likely negative or unknown samples have been classified.

- PEBL [Yu, Han, and Chang, 2002, Yu, Man, and Chang, 2004], Roc-SVM [Li and Liu, 2003], and both Peng, Zuo, and He [2008], and Zhang and Zuo [2009] use an iterative SVM to create their final classifier.

- MCLS in Chaudhari and Shevade [2012] uses a variant of this, an iterative, nonlinear least square SVM (LS-SVM). As mentioned above, unlike the standard iterative SVM, unlabeled samples classified as strongly positive are added to the probable positive set in addition to the classified negative samples being added to the probable negative set as in the other algorithms.

- The C-CRNE algorithm [Liu and Peng, 2014] takes the iterative SVM a step farther, selecting the final positive and negative sets using a weighted voting method called WVC (Weighted Voting Classifier) [Peng, Zuo, and He, 2008] over all SVM iterations.

Naïve Bayes using Expectation Maximization

The naïve Bayes algorithm was described in depth in Section 4.1.2. By itself, naïve Bayes (NB) is a probabilistic classifier that is used with fully labeled data. By using the Expectation Maximization algorithm (EM), missing data such as samples with missing labels can be incorporated into the classifier commonly abbreviated EM-NB. The EM algorithm is used to estimate parameters of statistical models when the models depend on unobserved latent variables—in this case missing labels. The EM algorithm is an iterative algorithm made up of two repeating steps: first the expectation step calculates the expectation of the log-likelihood using the current parameter estimate, then the maximization step computes new parameter estimates by maximizing the expected log-likelihood found in the expectation step.

- The Spy-EM or S-EM algorithm in Liu et al. [2002] and the PE_PUC from Yu and Li [2007] both use this EM-NB combination, as described above.

- An early weighted clustering PU algorithm [Jeon and Landgrebe, 1999] used the EM algorithm, but with a Gaussian Mixture Model classifier rather than naïve Bayes.

- The LGN algorithm from Li, Liu, and Ng [2007] goes in a different direction when expecting few negatives in an outlier detection situation. LGN constructs a single negative document (see Section 4.2.1 above) and, with the known positives, it creates a standard supervised naïve Bayes algorithm and uses it to identify any negative samples in the unlabeled set.

Other Methods

While most two-step PU learning algorithms use one of the above algorithms to create their final classifier, other methods are explored here.

- The PNLH algorithm [Fung et al., 2006] uses an iterative k-means to identify which samples are significantly similar to either the positive or negative by partitioning the dataset and selecting only those datapoints that are in the same cluster, or partition, as exclusively positive or negative samples. These are labeled appropriately; any ambiguous examples remain unlabeled. This process is repeated until no unlabeled samples remain.

- The categorical Pulse algorithm [Ienco and Pensa, 2016] builds two models, one for each class, using DILCA (*Distance Learning for Categorical Attributes* from Ienco, Pensa, and Meo [2012]). A modified kNN (k Nearest Neighbor) algorithm is used to place each unlabeled sample into either the positive or negative class based on DILCA distance and model features.

- The recent MMPU, or *Multi-Manifold PU learning* algorithm published in Gong et al. [2019b], also uses a unique classification solution. Here, unlabeled data are ranked along the learned manifolds to get confidence value of being positive. A confidence threshold is used to determine the positive/negative cutoff value.

4.2.3   SELECT TWO-STEP ALGORITHMS

Some of the more well-known two-step PU learning algorithms are described below. This is not intended to be an exhaustive list, but rather a closer look at some of the classic algorithms that are often referenced in literature reviews and by name in the field. Many more modern two-step algorithms are based on the ideas described in these papers. The algorithms are provided in chronological order of initial publication.

S-EM (Spy-EM) Developed by Liu et al. [2002]

One of the early algorithms in PU learning used a small validation set of known positive samples as unlabeled "spies" to help in the identification of probable negative samples. This technique was originally used for document classification using an ordered bag of words representation.

Step 1: To identify probable negative samples, the S-EM algorithm randomly removes a small percentage of known positive samples P and puts them with the unlabeled samples U as "spies" to make new sets P_{spy} and U_{spy}, respectively. A naïve Bayes (NB) algorithm is applied, and a probabilistic class label is assigned to each sample in U_{spy}. The retained spy samples in U_{spy} are then used to determine a threshold cutoff value which is used to determine which of the data is most likely to be negative.

Step 2: The known positive samples P, probable negative samples N_P, and the remaining unknown samples U are then classified using an iterative implementation of naïve Bayes using the expectation maximization (EM) algorithm. This continues until convergence.

PEBL (Positive Example Based Learning) Created by Yu, Han, and Chang [2002] and Yu, Man, and Chang [2004]

The PEBL algorithm, like many of the early PU algorithms was focused on text classification, particularly web-page classification. A later paper by Han et al. [2016] follows a very similar strategy but with the incorporation of an ensemble voting technique for final classifier selection. Ensemble methods will be described in more depth in Section 4.3.2.

Step 1: To identify probable negative samples, which they call *strong negatives*, PEBL uses a monotone disjunction list (or 1-DNF) algorithm to calculate which features occur most frequently in the known positive samples, identify which unlabeled samples also have those features, and remove them from the pool of possible negatives.

Step 2: The known positive samples P and probable negative samples N_P are then classified using an iterative SVM. Remaining unlabeled samples U are added into the P or N_P sets as appropriate if they fall beyond the margin and solidly into either class of the SVM. This process is repeated until all unknown samples have been classified.

Roc-SVM (Rocchio-Support Vector Machine) Developed by Li and Liu [2003]

This algorithm makes use of an older classification algorithm called the Rocchio classifier that was originally designed for information retrieval in the 1960s. Like the previous classifiers, the Roc-SVM algorithm was designed for text classification.

Step 1: To identify probable negative samples, the unlabeled set U is treated as negative which, along with the positive set P, is used to build a Rocchio classifier. Rocchio classification is a *nearest centroid classifier* applied to text documents that are represented by TFIDF (*term frequency—inverse document frequency*) feature vectors. Unlabeled documents that are closer to the unlabeled centroid are labeled as probable negative samples. Additional variants are described that incorporate k-means clustering to improve negative selection.

Step 2: An iterative SVM algorithm is used for classification, using the probable negative and known positive document samples. Documents classified as negative are added to the pool of probable negatives on the next iteration.

A-EM (Augmented EM) Developed by Li and Liu [2005]

The A-EM algorithm proposes to handle scenarios when the labeled positive and unlabeled positive distributions are different, leading to increased complexity. Text documents once again are the focus of this algorithm. Unlike most two-step algorithms, A-EM works not by identifying probable negative samples, but by adding large quantities of probable negative samples to the unknown set.

Step 1: A classifier like Naïve Bayes will have a difficult time separating the positive and negative classes if there are many unlabeled positive documents in the unlabeled set U. To handle this, the authors augment the unlabeled set with a large selection of irrelevant documents that are expected to be negative to increase the ratio of N/P in U.

Step 2: With an unlabeled set that is mostly negative, a Naïve Bayes classifier is learned using the Expectation Maximization algorithm (NB EM). The authors describe a method of estimating the f-score of each iteration using word counts. They propose selecting the final classifier based on whichever iteration of the EM algorithm provides the last ratio of f-score estimates greater than one.

PSoL (Positive Sample only Learning) by Wang et al. [2006]

Unlike the text document focus of the previously described two-step papers, the PSol algorithm was first introduced in the journal Bioinformatics as an algorithm for finding non-coding RNA genes. This, and other biological applications, have been some of the most motivating applications for PU learning after the original push for text classification.

Step 1: To identify probable negative samples, a *maximum distance minimum redundancy negative set* method is used. This identifies datapoints x_i from U that both maximizes the Euclidean distance between x_i and both the nearest known positive sample and other data points already added to the probable negative set N_P. This second condition is used to reduce the redundancy of the negative set.

Step 2: An iterative SVM algorithm is used for classification, using the probable negative and known positive document samples. New probable negatives are added to N_P after each iteration.

4.3   OTHER METHODS AND CONSIDERATIONS

While most PU learning algorithms either treat all unlabeled data as some sort of negative proxy or use the two-step algorithmic approach described in the previous section, some algorithms use other, unique solutions to the problem. In this section, we discuss some of these less common solutions to PU learning.

4.3.1   NEW LEARNING DIRECTIONS

Traditional PU learning methods have not usually considered scalability issues such as computational efficiency and overfitting, but this is a growing problem as datasets become ever larger. Some earlier work by du Plessis, Niu, and Sugiyama [2015] and Kiryo et al. [2017], discussed in depth in Section 4.1.4, began to look at scalability and computational efficiency. The USMO, or *Unlabeled data in Sequential Minimal Optimization*, algorithm by Sansone, De Natale, and Zhou [2019] starts with the convex optimization problem in du Plessis, Niu, and Sugiyama [2015] and uses Gram matrices to derive an algorithm that has much lower memory and computational requirements. Another paper, Kwon et al. [2020], proposes a solution for large-scale datasets by producing a closed form classifier under certain conditions. The authors build on the work of Sriperumbudur et al. [2012] for supervised learning that provided a closed-form classifier in supervised binary classification.

Other recent work has been focusing more on the structure of the data and data distributions. In addition to the recent two-step manifold solution proposed by Gong et al. [2019b], a paper by Gong et al. [2019a] uses the hat loss and a calibration term to attempt to learn additional information from the structure of the underlying data distributions. They call their method *Large-margin Label-calibrated SVM*, or LLSVM. Ke et al. [2018] uses a combination of global noisy negative learners (as discussed in Section 4.1.3) with local confirmation requiring that two samples belong in the same class only if they have a similar intrinsic geometry.

A promising new direction in PU learning is the incorporation of reinforcement learning techniques such as GANs (*Generative Adversarial Networks*) and Autoencoders. Unlike the algorithms in the previous paragraph, these methods are computationally expensive but extremely powerful, particularly for image classification and image processing. The most complex of these is GenPU (or *Generative Adversarial Positive-Unlabeled Learning*) published by Hou et al. [2018]. GenPU uses five deep learning neural networks—two generative and three discriminative. The two generative agents generate positive and negative samples respectively. The three discriminator agents are each trained to recognize one of the classes: positive, unlabeled, and negative. The authors provide a theoretical analysis that they claim GenPU can recover both positive and negative data distributions at equilibrium. General structure shown in Figure 4.6a. PGAN, or *Positive-GAN*, is another GAN based algorithm also published in 2018 by Chiaroni et al. [2018], intended for image classification. It is a two-step algorithm but is included in this section due to its use of reinforcement learning methods. The PGAN algorithm uses three deep neural networks—a generator and discriminator in step 1 to generate fake negative images, and a classifier in step 2. A block diagram of PGAN is shown in Figure 4.6b.

Two new GAN-based PU learning models called a-GAN (a direct adaptation of GAN for PU learning) and PAN (*Predictive Adversarial Networks*) have been recently proposed by Hu et al. [2021]. While a-GAN is an adaptation of a standard GAN for PU learning, an additional discriminator, called a classifier $C(\cdot)$, replaces the normal generator to try and distinguish likely positive samples from the unlabeled set U. The PAN model modifies a-GAN slightly and uses

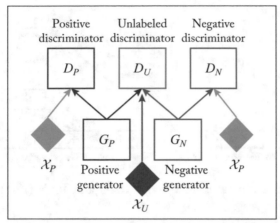

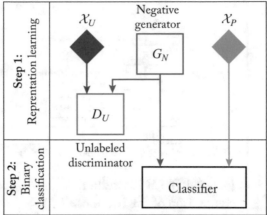

(a) GenPU with five deep neural networks   (b) PGAN with three deep neural networks

Figure 4.6: Two deep learn GAN-based PU learning algorithms. Images reproduced from Hou et al. [2018] (a) and Chiaroni et al. [2018] (b).

the sum of the KL-divergences of the predictions of all examples of the unlabeled set as the distance. D_i and C_i are used to denote the known positive and the unlabeled distributions and the KL-divergence is used to measure the distance between them as shown in the PAN objective distance function defined in Equation (4.26) from Hu et al. [2021].

$$
\min_{C} \max_{D} V(D, C) = - \underbrace{\sum_{i=1}^{n} KL(P_i^{pu} || D_i^{pu})}_{Term\ 1}
$$

$$
+ \lambda \left(\underbrace{\sum_{i=1}^{n_u} KL(D_i^u || C_i^u)}_{Term\ 2} - \underbrace{\sum_{i=1}^{n_u} KL(D_i^u || \tilde{C}_i^u)}_{Term\ 3} \right). \tag{4.26}
$$

The structures of these two new GANs models are illustrated in Figure 4.7.

In an interesting twist on using reinforcement learning for PU learning is called *Positive-Unlabeled Reward Learning*, or PURL by Xu and Denil [2019]. In this paper, the 2017 PU learning algorithm by Kiryo et al. [2017] is used to assist in a reinforcement learning problem—to help the discriminator suppress unreliable predicted awards. In 2020, a paper tying PU learning and open set domain adaptation together trains an autoencoder to reconstruct the known

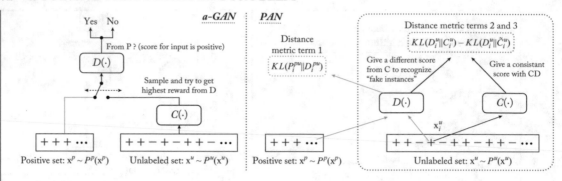

Figure 4.7: An illustration of the objective functions of a-GAN (left) and PAN (right) as a comparison of the two models.

samples and map the unknown samples before using reinforcement learning for source/target discrimination [Loghmani, Vincze, and Tommasi, 2020].

4.3.2   ENSEMBLE TECHNIQUES

Ensemble learning is a method that uses and combines multiple models to solve a problem, in our case a classification problem. Ensemble techniques can be especially helpful if the problem is sensitive to initial conditions or noise in the data. In this scenario, combining many weak learners can create a strong learner and reduce the likelihood of getting a poor classifier. As discussed in Section 2.4.2, the PU learning problem can be represented as a class-dependent noisy learning problem, making ensemble methods especially relevant in this field.

While there are many different ensemble methods in use today, the two most common in PU learning are bootstrap aggregating, usually abbreviated as bagging [Breiman, 1996] and shown in Figure 4.8, and iteratively built SVMs combined into an ensemble learner using a weighted voting scheme. With bagging, the data is subsampled into randomly chosen subsets with replacement allowed, meaning that the same sample can appear in multiple randomly drawn subsets. A learning algorithm is applied to each subset of data and the result, the positive or negative classification in this case, is remembered. After some number of subsets have been classified, the data's final output class is selected by voting over their bagged labels.

Perhaps the most cited ensemble PU learning algorithm is a 2014 paper called "A bagging SVM to learn from positive and unlabeled examples" [Mordelet and Vert, 2014]. In this paper, a straightforward bagging algorithm using SVM learners is described, the only difference being that the known positive elements are not subsampled but included in each subset as given—only the unlabeled data is randomly sampled and treated as negative by the classifier. The reasons given for this are that (a) the labeled positive data is known to be positive and (b) the number of labeled positive data is often limited. A standard SVM is used as the classifier for each bootstrap subset. Two hyperparameters are described—one for the size of each subset and one for the number of

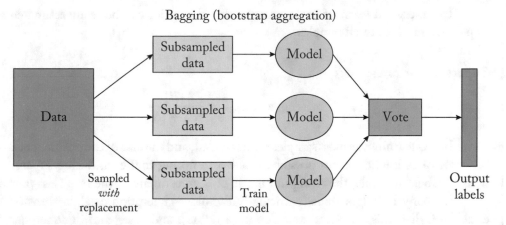

Figure 4.8: Bootstrap Aggregation, more commonly known as "Bagging," is an ensemble classification method that breaks the data into several smaller subsamples, typically allowing replacement, learns models on the subsamples, and then combines the models through voting to obtain a stronger final classifier.

subsets or bootstrap samples required. Mordelet and Vert first used an algorithm similar to this in Mordelet and Vert [2011], a gene prioritization algorithm more fully described in the next chapter. Claesen et al. [2015a] continued the work done in Mordelet and Vert [2014] with some modifications such as subsampling from the labeled positives and using class weighted SVMs to allow for a greater misclassification penalty for misclassified positives.

Several of the two-step algorithms described in Section 4.2 used a simple type of ensemble learner by saving either the data labels [Han et al., 2016] or the classifiers themselves [Liu and Peng, 2014, Peng, Zuo, and He, 2008] and combining them using a weighted voting scheme at the end of step two, after the iterative SVM has finished processing all of the data.

4.4   ESTIMATING THE CLASS PRIOR

In a classification scenario, as briefly discussed in Section 1.3, a binary classifier is related to the likelihood and the class prior (or prior probability) through Bayes' theorem.

Bayes' theorem

$$p\left(y = 1 | \boldsymbol{x}\right) = \frac{p(\boldsymbol{x} | y = 1) \cdot p(y = 1)}{p(\boldsymbol{x})}. \tag{4.27}$$

Here, the class prior $p(y = 1)$ (sometimes abbreviated as π) is the probability that a randomly chosen sample belongs to the positive class. If we can accurately estimate this, we can identify how many of the datapoints in our dataset are positive.

In supervised learning environments, this class prior can be easily estimated as the ratio of positive samples to all samples.

Supervised prior estimate

$$p(y = 1)_{est} = \frac{|P|}{|P| + |N|}.$$

(4.28)

In PU learning, as most samples are unlabeled, and the size of the positive set is not known, this type of estimation is not possible. It is worth noting that even during supervised learning with all labels available, the class prior is still only an estimate of the true class prevalence.

In many PU algorithms, particularly the older generation of algorithms such as Denis et al. [2003], Denis, Gilleron, and Tommasi [2002], and Garg and Sundararajan [2009], an accurate estimate of the class prior was required as an input hyperparameter to the system. This typically required a subject matter expert to manually estimate the prior using domain knowledge of the problem, which is limiting and can be unreliable. In one of the earliest PU algorithms, De Comité et al. [1999] suggested deviating from the purely PU nature of the data and extracting a small fully labeled dataset for class prior estimation. Other algorithms used relatively naïve estimators for the class prior such as Calvo [2008], which among others, estimates $p(y = 1)$ as the ratio of positives to unlabeled samples $|P|/|U|$. This is only effective if the unlabeled set mostly contains negative samples and is much larger than the positive set. More recent algorithms estimate the class prior more effectively as part of the learning process using the PU dataset itself. We will briefly describe some of these methods below and provide an in-depth overview of selected algorithms.

4.4.1   USING THE LABELING FREQUENCY C

The class prior $\pi = p(y = 1)$ is related to, but not the same as, the constant labeling frequency c from the SCAR assumption and described in Section 2.3.2. If either π or c is known, the other can be effectively estimated. The class prior describes the unconditional probability that a sample in the dataset is positive. This is a constant value as it does not depend on any of the feature values or the data distribution. Recall that c defines the probability that a positive sample will be *labeled* positive. Under the SCAR assumption, c is assumed to be a constant [Elkan and Noto, 2008].

$$c = p(s = 1|x, y = 1) = p(s = 1|y = 1).$$

(4.29)

The class prior and c are closely related. This can be seen if we apply Bayes law (shown in its general form in Table 1.1):

$$c = p(s = 1|y = 1) = \frac{p(y = 1|s = 1) \cdot p(s = 1)}{p(y = 1)}.$$

(4.30)

By rearranging terms, we can see that

$$p(y = 1) = \frac{p(y = 1|s = 1) \cdot p(s = 1)}{c}. \tag{4.31}$$

As only positive samples are ever labeled in the PU learning problem, if a data sample is labeled positive then the probability that it is positive is 100%. Therefore,

$$p(y = 1|s = 1) = 1. \tag{4.32}$$

Which allows us to simplify Equation (4.31) to show that under the SCAR assumption, the class prior equals

$$p(y = 1) = \frac{p(s = 1)}{c}. \tag{4.33}$$

The value $p(s = 1)$, which is technically a class prior as well, though this time of the class of *labeled* data samples rather than the positive or negative classes, can be estimated directly from the PU dataset as the number of labeled samples divided by number of all data samples m.

$$p(s = 1) = \frac{|P_L|}{|P_L| + |P_{UL}| + |N|} = \frac{|P_L|}{|P| + |N|} = \frac{1}{m} \cdot \sum_{i=1}^{m} s_i. \tag{4.34}$$

Using the relationship shown in Equation (4.33), the problem of estimating $\pi = p(y = 1)$ is reduced to the problem of estimating c. This can be done either by learning a non-traditional classifier $p(s = 1|x)$ as shown in Section 4.1.5 and described by Jaskie, Elkan, and Spanias [2019] or by using the top-down decision tree induction algorithm TIcE (*Tree Induction for c Estimation*) [Bekker and Davis, 2018a].

As mentioned in Section 2.3, the true class prior $p(y = 1)$ can only be guaranteed to be calculable using the single training-set data collection scenario. As Bekker and Davis [2020] demonstrates, we can, however, define a pseudo prior $\alpha = p(y = 1|s = 0)$ in the case-control scenario which lets us calculate the true class prior as

$$\pi = p(y = 1) = p(y = 1|s = 0) \cdot p(s = 0) + p(y = 1|s = 1) \cdot p(s = 1)$$
$$= \alpha \cdot p(s = 0) + p(s = 1). \tag{4.35}$$

By combining Equations (4.33) and (4.35), we find the conversions listed in the second part of Table 4.4.

4.4.2 USING MIXTURE MODELS AND F-DIVERGENCES

A mixture model is a probabilistic model that is used to represent subpopulations within an overall population. A quite complicated distribution can be modeled as a combination, or mixture, of simpler models. While any types of distributions can be used, Gaussian distributions are

Table 4.4: Summary of class prior conversions from the labeling frequency c

Single Training Set Scenario	$\pi = p(y = 1) = \dfrac{p(s = 1)}{c}$	(4.36)
	$c = \dfrac{p(s = 1)}{p(y = 1)}$	(4.37)
Case-Control Scenario	$\alpha = p(y = 1 \mid s = 0) = \dfrac{1 - c}{c} \cdot \dfrac{p(s = 1)}{1 - p(s = 1)}$	(4.38)
	$c = \dfrac{p(s = 1)}{\alpha \cdot p(s = 0) + p(s = 1)}$	(4.39)

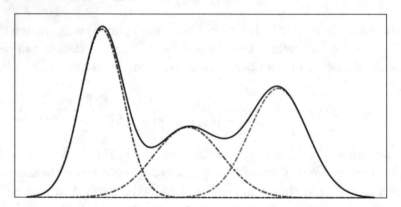

Figure 4.9: The black line shows a distribution composed of a mixture of three Gaussian sub-distributions (shown as dashed colored lines). This distribution could be modeled as a three component Gaussian Mixture Model (GMM).

the most common resulting in Gaussian Mixture Models, or GMMs. A GMM is a parametric probability density function that is represented as a weighted sum of Gaussian component densities [Reynolds, 2015]. GMMs are sometimes used for clustering or even for classification, though their main purpose is ultimately to describe a distribution of data. A simple binary classification problem can be thought of as a two-component mixture model, though as the positive and negative distributions become more complex, more components may be needed to fully capture that complexity. Mixture Proportion Estimation (or MPE) is one name for the problem of estimating the weight of each component in a multi-component mixture model given samples from both mixture and component.

A f-divergence is a function that measures the *difference* between two probability distributions. This can be used for MPE to learn a model of a distribution by minimizing the f-divergence between the model and the actual distribution from the existing data.

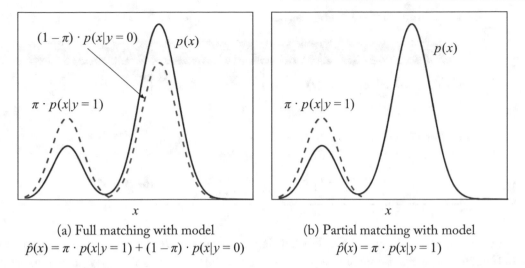

(a) Full matching with model
$$\hat{p}(x) = \pi \cdot p(x|y=1) + (1-\pi) \cdot p(x|y=0)$$

(b) Partial matching with model
$$\hat{p}(x) = \pi \cdot p(x|y=1)$$

Figure 4.10: Class-prior estimation by full matching (a) or partial matching (b) to unlabeled input data density $p(x)$. Recall that $\pi = p(y=1)$.

As mentioned in Section 2.2.2, in PU learning, the distribution of a datapoint x can be thought of as a two-component mixture model:

$$p(x) = p(y=1)\, p(x|y=1) + p(y=0)\, p(x|y=0). \tag{4.40}$$

du Plessis and Sugiyama [2014b] used the Kullback–Leibler divergence to learn the class balance for semi-supervised learning problems with some positive and some negative data. The same authors applied the Pearson divergence using partial matching to the PU learning problem using the SCAR assumption [du Plessis and Sugiyama, 2014a]. They argued that while full matching (Figure 4.10) is not possible without known negative data, if the positive and negative classes are not strongly overlapping, partial mapping, which is possible, should be effective. To do this, the divergence from $p(y=1)p(x|y=1)$ to $p(x)$ minimized. This is the method is used in Kanehira and Harada [2016].

du Plessis, Niu, and Sugiyama [2015a] found that applying the Pearson divergence consistently overestimated the true class prior. They proposed an alternate method using the penalized L1 distance that dramatically simplified the estimation procedure and obtained a better estimate of the class prior.

In 2016, three important papers were published on estimating the class prior for PU learning. Jain et al. [2016] described a nonparametric AlphaMax algorithm that uses the Kullback–Leibler divergence to estimate the mixing proportion of the two-component mixture model shown in Equation (4.40). Later that year, the same authors published another paper [Jain, White, and Radivojac, 2016] that is specifically robust to noise in the positive labeled data and in high dimensional feature spaces. They do this by modeling both the labeled positive distri-

Table 4.5: Common f-divergences (taken from du Plessis, Niu, and Sugiyama [2015a])

Divergence	Function $f(t)$	Conjugate $f^*(z)$	Penalized Conjugate $\tilde{f}^*(z)$
Kullback-Leibler	$-\log(t)$	$-\log(-z) - 1$	$\begin{cases} -1 - \log(-z) & z \leq -1 \\ z & z > -1 \end{cases}$
Pearson	$\frac{1}{2}(t-1)^2$	$\frac{1}{2}z^2 + z$	$\begin{cases} -1/2 & z < -1 \\ (1/2)z^2 + z & -1 \leq z \leq 0 \\ z & z > 0 \end{cases}$
L_1-distance	$\|t-1\|$	$\begin{cases} z & -1 \leq z \leq 1 \\ \infty & otherwise \end{cases}$	$\max(z, -1)$

bution $p(x|y = 1)$ and the unlabeled data distribution $p(x|s = 0)$ each as a two component GMM that share the same components but not the same weights. They present an extension of the EM algorithm that simultaneously estimates the parameters of both mixture models, taking advantage of the shared components.

Also in 2016, Ramaswamy, Scott, and Tewari estimated the proportions of a mixture model using an embedding of the distributions into a reproducing kernel Hilbert space. This is a very efficient algorithm as it requires only a simple quadratic programming solver.

4.4.3   OTHER METHODS

While the above class prior estimation methods are described primarily for the single training set data acquisition scenario, Kato et al. [2018] has attempted to estimate the class prior using the case control scenario specifically. Their solution is to alternate between estimating the class-prior and creating a PU classifier. This is done in two repeating steps until the prior estimation converges as follows.

Step 1: Given an estimated class-prior π^*, train a classifier f that minimizes the empirical risk $R^{\pi^*}(f)$:

$$R^{\pi^*}(f) = -\pi^* \mathbb{E}_1 [\log(f(X))] \\ + \pi^* \mathbb{E}_1 [\log(1 - f(X))] - \mathbb{E}_X [\log(1 - f(X))], \tag{4.41}$$

where $\mathbb{E}_1$ and $\mathbb{E}_X$ are the expectations over $p(x|y = 1)$ and $p(x)$, respectively.

Step 2: Denote the minimizer classifier from step 1 as f^*. Treat f^* as an approximation of $p(y = 1|x)$ and update π^* by taking the expectation of the classifier f^* over unlabeled data.

4.5   SUMMARY

In many ways, Chapter 4 lies at the heart of this book on PU learning. In this chapter we have provided an in-depth survey of many of the PU solutions that have been published since the field began. We have attempted to strike a balance between providing sufficient information to understand the nature and key components of each solution without detailing each step of every algorithm. The references provided encourage further in-depth investigation in any algorithm of interest.

CHAPTER 5

Applications

Just as supervised classification is used in many fields and for a variety of purposes, PU learning is extremely versatile and can be used in almost every application area of ML that involves classification. PU learning is used for automatic label identification in poorly supervised learning problems. In some areas, such as the biomedical field, PU learning is a natural fit as obtaining negative samples can be difficult or even impossible in some cases. In this chapter we present a survey of some existing PU applications to both illustrate work that is being done in this field and to encourage new ideas and directions for further research. These applications have been broken into broad, occasionally overlapping, categories and will be presented in alphabetical order in the sections below.

We provide references to papers that have included these topics as examples, with relative emphasis indicated. If a use-case is only mentioned in passing it will be given as a standard reference, if is reasonably well discussed or in the paper's experiments and results sections, it will be marked with an asterisk (*), and if that use case is the entire purpose of their paper, it will be marked with two asterisks (**). Public datasets referred to in these papers can be found in Appendix D.

5.1 ANOMALY OR OUTLIER DETECTION

Anomaly detection, including manufacturing fault detection, novelty detection, and outlier detection, is an important area in PU learning and is closely related to one class classification (OCC) which was described in some depth in Section 2.4.1. Abnormal events such as earthquakes or unusual medical events can also be considered anomalous. Many areas of manufacturing are concerned with identifying faulty products before shipment. Identifying microscopic faults in airline wings using sensors for flex analysis could improve airline safety.

- PU learning for anomaly/novelty detection has been extensively studied. The authors of Kato, Teshima, and Honda [2019], Kwon et al. [2020], and Hou et al. [2018] discuss the subject while Zhang et al. [2019]*, and especially Blanchard, Lee, and Scott [2010]** and Zhang et al. [2017]** delve deeply into the subject.

- Inlier-based outlier detection, with the goal of detecting outliers from an unlabeled set, given inlier samples, is investigated by Kato, Teshima, and Honda [2019], du Plessis, Niu, and Sugiyama [2014], and Sansone, De Natale, and Zhou [2019].

- Abnormal event detection is an area of great promise for PU learning. From the subtle signs of early earthquake detection to detecting the precursors of a seizure to allow the patient to prepare, many abnormal events have the potential to be detected early in time for advance preparation [Ren et al., 2018].

5.2 BIOMEDICAL APPLICATIONS

Biomedical applications are one of the most significant usage areas in PU learning due to their importance in our society, inherent cost, and fundamentally PU nature. By a fundamental PU nature, we mean that in biomedical problems it is much easier to detect the presence of a virus, a disease, or a drug interaction than it is to detect its absence. This makes these problems a very natural fit in PU learning.

5.2.1 GENETIC/BIOLOGICAL IDENTIFICATION AND ANALYSIS

Genetic identification was one of the earliest non-text uses of PU learning. In 2006, PSoL, an algorithm for finding non-coding RNA genes, was introduced by Wang et al. [2006]. More papers flooded in and have made this one of the most important applications of PU learning to this day. Unlike other fields such as computer vision, text classification, remote sensing, etc., it is not just *expensive* to label negative samples in this field, but it is often *impossible*. While it is possible to identify some genes as influencing a disease such as Alzheimer's disease, cystic fibrosis, sickle cell anemia, or even anxiety disorders and other diseases, it is impossible, or nearly impossible with our current technology, to determine that a certain gene does NOT influence the disease. ***Absence of evidence is not evidence of absence.*** This makes these problems inherently difficult to solve but possible using PU learning.

There are many different variants of this type of genetic/biological identification and analysis problem in the literature. These are summarized below.

Gene Identification

- Disease gene identification—identifying genes that contribute to or influence a disease with a genetic component—is one of the most important biomedical applications of PU learning. Some diseases have genetic components that a person is born with, such as cystic fibrosis, while other diseases such as cancer involve genetic mutations. Work in this field has focused on identifying new genes that influence genetic diseases [Jaskie, Elkan, and Spanias, 2019, Kwon et al., 2020, Mordelet and Vert, 2014], [Mordelet and Vert, 2011]**, [Calvo, 2008, Yang et al., 2012, 2014]**, and [Geurts, 2011, Teisseyre, Mielniczuk, and Łazecka, 2020, Xu et al., 2017, Yang, Liu, and Yang, 2017]. However, Calvo [2008]** also looks at genes involved in the cancer process.

- It is not only disease genes that are worth identifying. Many other types are genes are identified and analyzed using PU learning [Mordelet and Vert, 2014]*, [Claesen et al.,

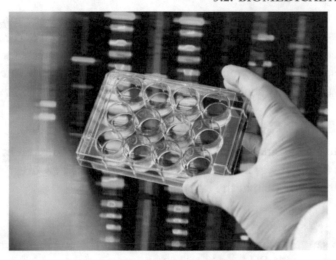

Figure 5.1: Genetic research wet work.

2015a, Pham and Raich, 2018]. An example of this is found in Cerulo, Elkan, and Ceccarelli [2010]** where PU learning is used to reconstruct gene regulatory networks from gene expression data. Two genes are predicted to interact if the correlation coefficient of their expression levels is above a set threshold. Another example from Wang et al. [2006]** uses PU learning to identify small non-coding RNA (ncRNA) genes from intergenic sequences.

- Other genetic PU learning applications include gene database completion [Bekker and Davis, 2018a, Geurts, 2011] and bioinformatics classification in general [Ke et al., 2018, Ren et al., 2018].

Proteins and Protein Functions
- Another biomedical application involves identifying transport proteins for a specialized Transportation Classification Database (TCDB) from the general protein database SwissProt. This is a useful problem because it demonstrates the ability to construct many smaller, custom, biological (or other) datasets from large general-purpose ones [Das, Saier, and Elkan, 2007]**, [Elkan and Noto, 2008, He et al., 2018]*.

- Protein function assignment is another use-case. For example, most enzymes are proteins, but not all proteins are enzymes [Jain et al., 2016, Jain, White, and Radivojac, 2016, Ramola, Jain, and Radivojac, 2019]. Predicting protein-protein activity [Jain, White, and Radivojac, 2017] or interactions [Geurts, 2011] and protein pupylation sites [Nan et al., 2017]** are further biomedical applications.

- An interesting application, protein structure determination using cryo-electron microscopy (cryoEM), is a specific application for a type of computer vision problem related to the biomedical field. Identifying individual particles (or "particle picking") in images can require months of manual effort but can be reduced to seconds or minutes using PU learning [Bepler et al., 2019]**.

5.2.2 HEALTHCARE AND DRUG DEVELOPMENT

In addition to genetic analysis, healthcare is an area of great interest in PU learning. Can health and diseases management be improved? Can we discover new drug treatments or interactions on a computer rather than in the laboratory?

Healthcare

- We saw that disease gene identification in the previous section is an important area of research in PU learning, but another area of equal interest is the idea of identifying diseases by looking through healthcare records and databases. Diseases that can be difficult to diagnose early, such as cancer or diabetes, may be identifiable by examining blood test results and other patient history records. Many authors have suggested that this problem is a good fit for PU learning [Bekker and Davis, 2018a,b, 2019, De Comité et al., 1999, Denis, 1998, Denis, Gilleron, and Letouzey, 2005, Pham and Raich, 2018, Yu, 2005, Zhang and Lee, 2005].

- Recent work is being done in predicting hospital length of stay by Arjannikov and Tzanetakis [2021]** and Arjannikov [2021]**.

- In 2019, Reamaroon et al. [2019]** applied this method to the Acute Respiratory Distress Syndrome (ARDS), a critical illness syndrome that affects 200,000 people a year in the United States with a mortality rate of 30%, but is vastly under-recognized.

- Another application in healthcare is the prediction of surgical site infections (SSIs) using features such as wound edge distance, wound edge color, patient heart rate, blood pressure, and so on. Ren et al. [2018]* was able to construct an effective classifier to predict SSIs in the first nine days after surgery.

- Using PU learning to automatically classify vascular legions from medical images is an effective and inexpensive solution for an important task in the diagnosis and follow-up of coronary heart disease [Zuluaga et al., 2011]** [Ienco and Pensa, 2016].

- An important problem and natural fit for PU learning is for biomedical rare event detection and labeling. Ren et al. [2018]* has used this for EEG seizure detection and other potential applications include ECG heart arrhythmia or heart attack detection. Begum et al. [2013]** explores PU learning for abnormal heart arrhythmias along

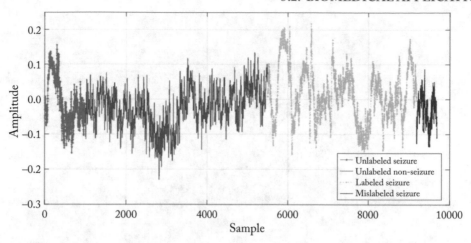

Figure 5.2: EEG signal with yellow and blue seizure signals and orange and purple non-seizure signals. Yellow seizure section is correctly labeled positive, purple is mis-labeled positive, and orange and blue are unlabeled. From Ren et al. [2018].

with identifying heartbeats associated to sudden cardiac death from a Sudden Cardiac Death Holter Database.

Computational Drug Discovery

- Given a large chemical database, virtual drug screening, also called drug repositioning, is the identification of existing drugs to be used for new uses and to help with new diseases or disease symptoms can be much more time and cost effective than developing new drugs from scratch [Claesen et al., 2015a], [Liu et al., 2017]**. This is also a natural PU learning problem as only very limited positive interactions may be known.

- Investigating drug-drug interactions (DDIs) is a related and similarly important problem [Hameed et al., 2017, Liu et al., 2017]**.

5.2.3   ECOLOGICAL AND ENVIRONMENTAL MONITORING

PU learning for ecological and environmental monitoring are scenarios that have only been lightly touched upon in the literature but have excellent future potential.

- Of great interest in ecological and environmental research, identifying species population and presence can be a difficult problem. A shy species can be difficult to locate, while determining species absence in a region is expensive. Geographical regions with reported sightings of the species of interest make up the positive set, while all other areas remain in the unknown set. The authors of Ward et al. [2009]** and Li, Guo,

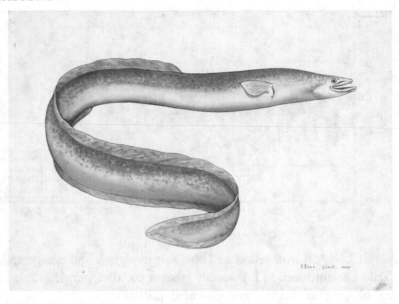

Figure 5.3: New Zealand Longfin Eel. **Artwork by Frank Olsen, Queensland State Archives, Item Representation ID DR6164.**

and Elkan [2011b]** use PU learning to model species presence. Ward et al. [2009]** provides a case study concerning species presence modeling of the Longfin Eel (shown in Figure 5.3) in the rivers of New Zealand. PU learning was able to identify the most important environmental factors for the eel's survival and presence.

- Environmental monitoring applications such as monitoring land-use change, measuring water quality, and mapping vegetation are remote sensing applications that can be solved using PU learning [Gan, Zhang, and Song, 2017, Li, Guo, and Elkan, 2011a]. Remote sensing will be described in more detail in Section 5.4.3.

5.2.4   OTHER BIOMEDICAL APPLICATIONS

The applications listed above form only a small percentage of the possible biomedical applications that could benefit from PU learning.

- An interesting medical problem, closely related to PU learning but more difficult—that of learning only from the proportions of the PU sets—is being used for embryo selection in assisted reproduction [Hernández-González, Inza, and Lozano, 2017]**. To maximize the chances of a healthy baby while minimizing the likelihood of giving birth to many children at once, only the most viable embryos should be implanted, but it is unknown which of the embryos ultimately implants in the woman's uterus. Thus, only the proportion of successes are known, the individual positive samples are not.

- The growing medical fields concerning gut microbiome analysis and epigenetics could also benefit from PU learning [Jaskie and Spanias, 2019].

5.3   BUSINESS APPLICATIONS

While all PU applications could have some type of ultimate business purpose, the applications described in this section are directly applied to marketing, market analysis, and social networking—all areas traditionally considered part of the business domain.

5.3.1   RECOMMENDER SYSTEMS AND DIRECT MARKETING

The financial value of recommender systems and direct marketing to the companies that use them is difficult to overstate. Recommender systems are today used by nearly all major websites today. Netflix states that their recommender systems (which recommend a customized list of shows for each person based on their predicted preferences) influence approximately 80% of what people watch. The combined effects of personalization and recommendation save them more than $1 billion per year [Gomez-Uribe and Hunt, 2015]. YouTube reports that 60% of its clicks are due to recommendations, and Amazon's CEO said in 2006 that 35% of their sales originated from cross-sales, meaning recommendations [Jannach and Jugovac, 2019].

- While modern recommender systems typically have both positive and negative feedback options (such as thumbs up and thumbs down buttons), it is often possible to determine if a person liked a video or movie even without direct feedback by determining if they watched it through to the end. Negative feedback in such a situation is more difficult as passing over a video or even leaving the video early may be due to other factors than dislike. This makes the recommender problem especially well-suited for PU learning, as shown in Chang et al. [2016]* and de Campos et al. [2018]**.

- Direct, or targeted, marketing allows a business to save a significant amount of money and is a natural match for PU learning. Lee and Liu [2003], Zhang and Lee [2005], and Denis [1998] suggest that a business could have known customer profiles compose the positive set and large unknown databases of customer information make up the negative set to identify potential new customers.

- Social network marketing is a type of direct marketing that uses social networks rather than traditional customer databases and is a perfect fit for PU learning as in general, only positive "likes" can be collected. The authors of Jain et al. [2016], Jain, White, and Radivojac [2017], and Jain, White, and Radivojac [2016] briefly address this topic.

5.3.2   FRAUD DETECTION

As our digital world increases in size and scope, digital fraud becomes an ever-greater problem. Digital fraud includes such disparate topics as identity theft and stolen credit card numbers to fake product reviews, email scams, and insurance fraud. This section has significant overlap with text classification (Section 5.5), anomaly or outlier classification (Section 5.1), cybersecurity (Section 5.6.2), and time series (Section 5.6.4) applications. It is included in its own business subsection due to its individual importance and associated financial costs.

- According to the Financial Cost of Fraud Report [Crowe, 2019], "globally, fraud losses equate to a shocking US\$5.127 trillion each year, which represents almost 70% of the \$7.442 trillion which world spends on healthcare each year." Many types of fraud detection are a natural fit for PU learning. It is often possible to identify *some* examples of fraud, fraudulent credit card transactions for example that may be confirmed with the card holder, but it is difficult to know that a transaction is NOT fraudulent without great expense and bother to the card holder. Fraud can be much easier to *confirm* than it is to *find* making this a natural fit for PU learning. Jannach and Jugovac [2019], Jaskie, Elkan, and Spanias [2019], Jaskie and Spanias [2019], and Chang et al. [2016] briefly discuss this problem but we believe there is room for growth in this area.

- A specific type of fraud, fake online product and business reviews, also has a substantial cost to our society. 65% of reviews on sites like Amazon, despite crack-downs on fraudulent product reviews, may be fake and the actual costs of bad business reputation caused by fake business reviews have been estimated to exceed \$500 billion in the U.S. alone [Sickler, 2018]. As it can be difficult to know for sure if a review is NOT fraudulent, this is another application that fits well with PU learning. Fortunately, this is an area that PU researchers have explored [Gong et al., 2019a] and investigated in some depth [Li et al., 2014, Ren, Ji, and Zhang, 2014]**.

5.3.3   OTHER BUSINESS APPLICATIONS

In addition to the business applications described above in some depth, there are many other business applications that have been addressed in the literature, as well as undoubtably, more to be thought of in the future. We attach a summary of some existing applications.

- Social network analysis and predicting of future connections between people and "liked" products or business can be thought of as a streaming PU learning application. The authors of Pham and Raich [2018] and especially Chang et al. [2016]* tackle this analysis and prediction task.

- Reject inference for loan approval and other tasks learns from the applications of both accepted and rejected individuals. The behavior of rejected individuals is unknown and thus fits within the Positive and Unlabeled framework. Smith and Elkan [2004]** applies PU learning to reject inference problems which can include epidemiology, econometrics, and clinical trial evaluation along with the more standard financial applications.

- When gathering survey data, socially stigmatized topics such as alcoholism, texting while driving, or diseases such as STDs are often under-reported. In Teisseyre, Mielniczuk, and Łazecka [2020], PU learning is discussed as a possible method for compensating for this user-reported bias.

- Opinion mining, also known as sentiment analysis, is a type of natural language processing (also described in Section 5.6.3) that extracts sentiment or opinion from within text (positive, negative, neutral, etc). Ke et al. [2018] suggest that this may be a good fit for PU learning though no details are provided.

5.4 COMPUTER VISION

Computer vision is an important and growing area in ML and AI and its applications in PU learning have been growing rapidly. One of the earliest and still most important PU computer vision applications is the field of remote sensing—identifying land types, targets, clouds, textures, etc., from imagery taken far above the earth's surface. In recent years, image and even video classification has emerged as PU learning applications.

5.4.1 IMAGE CLASSIFICATION

Image classification is the process of assigning a class label to an image. As PU classification is a binary classification problem, a common way of performing image classification using PU learning is to select one class as positive and all others as negative. This is sometimes called one class classification or one-vs.-all classification as described in Section 2.4.1.

- Image classification in general is described briefly in Jaskie, Elkan, and Spanias [2019], Li, Guo, and Elkan [2011a], Northcutt, Wu, and Chuang [2017], and Kwon et al. [2020] and in much more depth in Zhang et al. [2019]*, Gong et al. [2019b]*, Zhang et al. [2017]*, and most especially Chiaroni et al. [2018]**.

- Specific applications for image classification include automatic face identification or tagging [Kato, Teshima, and Honda, 2019, du Plessis, Niu, and Sugiyama, 2015], [Gong et al., 2019b]*, and even facial authentication which implies a certain level of confidence [Xu et al., 2017].

- Multi-label PU learning is a more complex forms of image classification where multiple items are labeled in a single image as shown in Figure 5.4 and described in Kanehira and Harada [2016]**.

Figure 5.4: Multi-label image: bird, cactus. Photo by author.

5.4.2 VIDEO CLASSIFICATION

Video classification is much more complex than image classification and often it is behaviors or activity in the video that is classified. Video classification is a computer vision application of a time series classification (described further in Section 5.6.4).

- Video activity categorization is an area of growing importance in our modern world. Identifying abnormal crowd behaviors [Zhang et al., 2017]* can help in riot and crowd control.

- Activity or emotion classification could detect nervousness or stress and assist in identifying persons to subject to more strenuous security checks at airports or large events. Illegal or dangerous activities such as an unexpected package being left in a populated area could result in a proactive police presence for increased safety [Jaskie, Elkan, and Spanias, 2019].

- Violent behavior detection can be identified quickly so that strategies and solutions can be rapidly deployed. Both Shi et al. [2018]* and Gong et al. [2019b]* test this application using a video HockeyFight dataset illustrated in Figure 5.5 (dataset details in Appendix D).

5.4.3 REMOTE SENSING

Remote sensing generally refers to the use of aerial imagery and sensor technologies, typically taken from satellites, airplanes, or drones, to acquire information about the Earth. It is used in many domains and includes commercial, military, economic, and scientific applications.

- While there are many different remote sensing applications, land-type classification is the most common PU remote sensing use-case in the literature. Land-type classification can be used to identify urban areas, identify shifting ecological boundaries,

Figure 5.5: Example video frames in the HockeyFight dataset from Gong et al. [2019b].

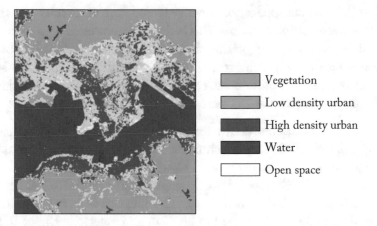

Figure 5.6: An example of multi-class PU land-type classification around the Victoria Harbor in Hong Kong, September, 2001 from Liu et al. [2006].

monitor land-use changes, map vegetation, the extent of floodwater boundaries (as shown in Figure 5.6), and so on. The authors Gong et al. [2019a], Hou et al. [2018], du Plessis, Niu, and Sugiyama [2015], and Xu et al. [2017] mention land-type classification in passing while Liu et al. [2006]* and especially Liu et al. [2018]**, Li, Guo, and Elkan [2011a]**, and Li, Guo, and Elkan [2020]** delve deeply into the subject.

• As another possible use, Jaskie and Spanias [2019] suggests that land-type classification could be used to identify man-made objects in satellite imagery such as new archeological sites or unknown military installations.

• Other applications for remote sensing include target or object detection/identification, cloud identification, and various texture detection and segmentation possibilities [Jeon and Landgrebe, 1999, Li, Guo, and Elkan, 2011a].

• It is also common, when performing remote sensing, to take advantage of more than the conventional three color RGB images. Hyperspectral images (HSI) often uses several spectral bands and can be used to increase the information available and the feature space [Gong et al., 2019a]. The authors of Gong et al. [2019b]* used a hyperspectral dataset known as Jasper Ridge in one of their experiments. This dataset is only 100×100 pixels in size, but each pixel has 198 spectral channels ranging from 380–2500 nm instead of the standard 3 for a RGB image. This means that while the 2D image size in pixels is small, the information contained in that image is more like a cube than a surface.

5.5   TEXT CLASSIFICATION

The earliest PU learning applications involved some form of text classification—often for website or document classification purposes. This included recommending new webpages and document or information retrieval, categorizing the subject of a paper, webpage, or email. One common application is email spam identification. Users identify some emails as spam, and these make up the positive class. All other emails are considered unknown. Text classification remains an important and prolific area of PU learning.

- Text classification usually refers to the idea of classifying documents by topic. This requires feature extraction techniques which vary by paper. Several papers such as Bekker and Davis [2018a], Ienco and Pensa [2016], Jaskie, Elkan, and Spanias [2019], Kwon et al. [2020], Teisseyre, Mielniczuk, and Łazecka [2020], Xu et al. [2017], Yu [2005], and Ren et al. [2018] mention text classification as an application in passing, but Zhang and Lee [2005]*, Lee and Liu [2003]*, and especially Liu et al. [2003]**, Li, Liu, and Ng [2010]**, Nigam et al. [2000]**, Denis, Gilleron, and Tommasi [2002]**, Liu et al. [2002]**, Li and Liu [2003]**, Han et al. [2016]**, Li and Liu [2005]**, and Denis et al. [2003]** are dedicated to text classification in particular.

- Website classification is like text classification, the only difference being that instead of documents, web pages or websites are used. This is separated from text classification only because a distinction between the two is often noted in the literature. While papers by Bekker and Davis [2018b], Claesen et al. [2015a], De Comité et al. [1999], Ienco and Pensa [2016], Ke et al. [2018], Lee and Liu [2003], Yu [2005], Zhang and Lee [2005], and Teisseyre, Mielniczuk, and Łazecka [2020] briefly comment on web page classification in passing, Blum and Mitchell [1998]*, Nigam et al. [2000]*, and particularly Yu, Han, and Chang [2002]**, Han et al. [2016]**, and Li and Liu [2005]** focus on the topic in detail.

- The related problem of networked text classification is described by Li et al. [2016]** and Chang et al. [2016]*.

- Information or data retrieval is a very important application of text or document classification. Information retrieval searches for information in a document, identifies the document of interest itself, finds metadata that describes the data of interest, and can be extended to databases of images, videos, sounds, and other media [Kato et al., 2018, Kato, Teshima, and Honda, 2019, Mordelet and Vert, 2014, Pham and Raich, 2018, Sansone, De Natale, and Zhou, 2019, Xu et al., 2017].

5.6   OTHER PU APPLICATIONS

This section includes some of the PU applications that don't fit well into one of the general categories discussed above.

5.6.1   ENERGY APPLICATIONS

Energy applications are a growing area in ML and PU learning. Several studies have made use of ML for solar energy monitoring and control. The authors of Rao et al. [2020] described several methods for fault detection and in fact an entire cyber-physical project with a testbed that uses sensors and machine learning is detailed in Rao et al. [2016]. Fault detection studies started in the beginning by clustering data using simple k-means algorithms clustering data on voltage current curves [Rao et al., 2016]. Several other ML algorithms for fault detection have also been reported in Garoudja et al. [2017] and Perera, Aung, and Woon [2014]. Gradually, these studies progressed to using neural network architectures [Chine et al., 2016, Li et al., 2021, Rao, Spanias, and Tepedelenlioglu, 2019] and more recently optimized pruned neural networks [Rao et al., 2020] and graph-based ML [Fan et al., 2020, Zhao et al., 2015]. Important studies on topology optimization have also been proposed where neural network architectures switch among standard PV topologies to optimize solar array output power [Narayanaswamy et al., 2019].

- PU learning innovations in fault detection have recently been described by Jaskie, Martin, and Spanias [2021]. In this study, PU learning was shown to be able to learn solar fault detection models using only a small set of labeled data relative to other ML methods. In the same study, a feedback enhanced PU learning algorithm was shown to improve detection accuracy with a small number of features.

- Cloud identification, which can be useful for modifying live solar panel connectivity [Narayanaswamy et al., 2019], is another PU application [Jeon and Landgrebe, 1999].

5.6.2   SECURITY APPLICATIONS

As the world becomes more connected and reliant on the internet, the internet of things (IoT), and computer systems in general, the danger of cyberattacks and other types of network intrusion become of increasing concern. PU learning can help with attack detection by using existing attack information to make up the positive class, while all other data is used to populate the unlabeled class.

- Network intrusion detection is explored in the experiment sections of Zhang et al. [2017, 2019]* and Qin et al. [2013]*.

5.6.3 SPEECH AND AUDIO PROCESSING

Our review of applications in speech and audio processing is organized by sub-area, namely, speech compression, speech and speaker recognition, and wideband audio processing. ML has been used in speech compression since the late 1970s with efforts to design codebooks using vector quantization for low rate coders in Buzo et al. [1979], Gray [1984], Makhoul, Roucos, and Gish [1985], Schroeder and Atal [1985], and Jaskie and Fette [1992]. In fact, the use of codebooks that quantized speech parameters at less than one bit per second enabled digital cell-phone telephony standards in the late 1980s [Spanias, 1994]. Some reference to semi-supervised ML methods for coding was made by Deng et al. [2010].

In terms of speech and speaker recognition early methods were based on Hidden Markov Models [Rabiner, 1989] and neural networks [Lippmann, 1989]. Logistic regression was applied to voice recognition in Birkenes et al. [2007] and PU learning for natural language processing was addressed in Li et al. [2010] which has shown that PU learning outperformed classical distributional similarity methods for expanding data sets for training. In all there are several examples of using semi-supervised learning for speech and audio processing and some of them are listed in Zhang et al. [2019] and Moreno and Agarwal [2003].

- Applications for PU speech and audio processing can include acoustic classification of audio signals [Jaskie, Elkan, and Spanias, 2019], as well as the very intriguing audio event detection described in Kumar and Raj [2016]**. The positive set is composed of a target sound or sounds. General audio from the target location makes up the unlabeled set. This can be used for diverse applications including songbird identification, gunfire detection, helicopter approach, or human presence detection.

- More closely related to Natural Language Processing (NLP), PU learning has been found helpful for set expansion [Li et al., 2010]**, learning word embeddings in a low-resource language [Jiang et al., 2018]**, and identifying word similarity and analogies and next word prediction [Tanielian and Vasile, 2019]*.

5.6.4 TIME SERIES ANALYSIS APPLICATIONS

Data stream applications, often called time series, streaming, or signal processing applications, are an important and growing area in ML in general. PU learning is frequently a good fit for these types of problems as often only items of interest are flagged in the signal. Many of the applications already described can be thought of as time-signal applications rather than static data classification problems. For example, credit card fraud is ideally detected live, as credit card transactions stream into the processing center at an enormous rate. Many biomedical applications such as detecting seizures or arrhythmias inherently involve signal processing as EEGs and

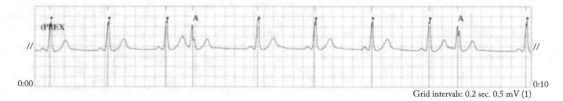

Figure 5.7: EKG data showing heart arrhythmias from the support website of Begum et al. [2013].

EKGs produce brain signals and heart signals, respectively. In this section we include papers that most directly focus on PU time-series algorithms with general or miscellaneous applications. Specific applications are gone into more depth in their respective sections.

- A growing subfield in PU learning is for time-series and signal processing PU algorithms specifically. The following papers provide a good resources in this area: Chang et al. [2016]**, Reamaroon et al. [2019]**, Qin et al. [2013]**, Liang et al. [2012]**, de Carvalho Pagliosa and de Mello [2018]**, Nguyen, Li, and Ng [2011]**, Li et al. [2009]**, González et al. [2016]**. Marussy and Buza [2013]** provides general semi-supervised time-series algorithms, and Begum et al. [2013]** introduces a PU learning algorithm that allows only a single labeled positive datapoint.

- Interesting time series applications were explored in the experiment sections of Ren et al. [2018]*, Begum et al. [2013]**, and de Carvalho Pagliosa and de Mello [2018]** including applications to population growth, atmospheric weather forecasting, sunspot identification, abnormal heart arrhythmias, surgical site infection, and seizure detection.

5.7   SUMMARY

In this chapter, we have attempted to provide a brief survey of some of the many and varied applications for which PU learning is being used. As with traditional classification, PU learning is a general-purpose learning problem with broad application and relevance to the modern world. Some problem types, such as many of the biomedical applications described in Section 5.2 would not be possible without weakening the labeling requirements of fully supervised learning. Other applications become faster, cheaper, and more effective by taking advantage of the reduced labeling requirements of PU learning.

As ML in general continues to expand and become ever more ubiquitous in our world today, new and exciting areas such as quantum ML provide potential future avenues to take advantage of PU learning's unique capabilities.

CHAPTER 6

Summary

6.1   CONCLUDING REMARKS

The Positive and Unlabeled learning problem, PU learning as we have called it throughout this book, is an area of growing importance in semi-supervised machine learning. In the modern era, enormous quantities of data are accumulated, yet labeling these data can be prohibitively expensive, time-consuming, and sometimes even impossible, making supervised learning algorithms unusable. Semi-supervised learning addresses those situations where all training labels are not known. Many real-world classification problems intrinsically have only a small set of known positive data along with large amounts of unlabeled or unknown data. In this book, we have attempted to define and describe the growing semi-supervised field of PU learning, provide a thorough survey of the algorithms and solutions created to solve it, and describe some part of the remarkable breadth of applications for which it is used. Our hope has been to help lay a foundation for study in this important field.

As this field has gained rapid popularity in recent years, the number and variety of scientific papers describing new solutions and applications is increasing exponentially. During the writing of this text, many new papers became available and while some were incorporated, many were not, or this text would never have been ready to be published. We look forward to the opportunity to incorporate these new ideas and applications in future editions.

6.2   FURTHER READING

Machine Learning is the science of finding patterns in big data. Many algorithms exist today, but most can be generally grouped in to one of three learning paradigms: supervised learning, unsupervised learning, and reinforcement learning. In this book, we have focused on the semi-supervised problem of Positive and Unlabeled learning but there are many varied topics in machine learning of equal interest. In this section, we provide some recommended resources for studying machine learning in general and suggest some niche areas of interest that look to grow into important new fields in their own right.

General Machine Learning books:

- One of the fundamental books in machine learning is *Pattern Recognition and Machine Learning* by Christopher Bishop [Bishop, 2011].

- *The Elements of Statistical Learning: Data Mining, Inference, and Prediction* by T. Hastie, R. Tibshirani, and J. Friedman describes the mathematical concepts behind ML algorithms [Hastie, Tibshirani, and Friedman, 2016].

- An older, but still useful, book for those getting started in ML is Tom Mitchell's classic *Machine Learning* [Mitchell, 1997].

Semi-Supervised Machine Learning:

- One of the few books entirely focused on semi-supervised ML, though lacking coverage in the PU learning problem specifically is *Semi-Supervised Learning* by O. Chapelle, B. Schölkopf, and A. Zien [Chapelle, Schölkopf, and Zien, 2006].

Deep learning has exploded in the last decade and many deep learning methods such as CNNs, RNNs, and GANs are being used in amazing and dramatic new ways in ML in general and in PU learning.

- The inventor of the GANs concept, Ian Goodfellow, along with Y. Bengio and A. Courville have a book called *Deep Learning* that comes highly recommended [Goodfellow, Bengio, and Courville, 2016].

An area of growing importance in ML is that of the Internet of Things (IoT), sensors, and "tiny ML" where machine learning algorithms, including PU learning, are done on microchips for what's called, Machine Learning on the Edge. Some great resources in this area include:

- *Sensors for IoT Applications* by M. Stanley and J. M. Lee [Stanley and Lee, 2018].

- The survey paper "A Brief Survey of Machine Learning Methods and their Sensor and IoT Applications" by U. Shanthamallu, A. Spanias, C. Tepedelenlioglu, and M. Stanley [Shanthamallu et al., 2017].

- A great new book in the rapidly growing community of TinyML is called *TinyML: Machine Learning with TensorFlow Lite on Arduino and Ultra-Low-Power Microcontrollers* by P. Warden and D. Situnayake [Warden and Situnayake, 2020].

Further websites, code, and video resources are available in Appendix E: Additional Resources.

APPENDIX A

Notation Used Throughout the Text

Table A.1: (*Continues*)

Symbol	Sections Defined	Definition
x	1.3, 2.2.1	A single data sample stored as a $1 \times n$ feature vector of n features.
y	1.3, 2.2.1	The label for x: $y = 1$ when x is positive and $y = 0$ when x is negative.
s	2.2.2	This variable represents whether a label is known or unknown in PU learning.
X	2.2.1	A $m \times n$ matrix of m data samples each with n features.
y	2.2.1	A $m \times 1$ column vector containing the labels for X.
m	2.2.1	The number of data samples in X.
n	2.2.1	The number of features in row vector x in X.
$f : \mathbb{R}^{1 \times n} \to [0, 1]$, $f(x)$, or f	2.2.1	A classification model that takes data x as input and predicts the probability that $y = 1$.
$\hat{x}$	2.2.1	A datapoint with unknown label, generally used for testing after model creation.
$\hat{y}$	2.2.1	The predicted label for datapoint $\hat{x}$.
P	2.2.2	The set of positive samples in X.
N	2.2.2	The set of negative samples in X.
U	2.2.2	The set of both positive and negative unlabeled samples in X when using PU learning. $U = P_{UL} \cup N_{UL}$

Table A.1: (*Continued*)

Symbol	Sections Defined	Definition
P_L	2.2.2	The set of labeled positive samples in X when using PU learning.
P_{UL}	2.2.2	The set of unlabeled positive samples in X when using PU learning. $P_{UL} = P - P_L$
N_{UL}	2.2.2	The set of unlabeled negative samples in X when using PU learning. $N_{UL} = N$
N_P	4.2.1	The set of probable negative samples in X when performing a two-step PU learning algorithm.
c	2.3.2	The label frequency – The unknown proportion of labeled positive samples out of all positive samples. $c = |P_L|/|P|$. This is often assumed to be constant using the SCAR assumption discussed in Section 2.4.1.
$p(y = 1)$	1.3, 2.2.2	The class prior. The probability that a random sample is positive.
π	2.2.2	An alternate notation for the class prior $p(y = 1)$.
$p(s = 1)$	2.3.2	The labeled prior. That is, the probability that a data sample is labeled.

APPENDIX B

Acronyms and Algorithms

Table B.1: (*Continues*)

Acronym	Meaning
1-DNF	Monotone disjunction list
AODE	Averaged one-dependence estimators
A-EM	Augmented EM
CNN	Convolutional neural network
C-CRNE	Clustering-basted method for collecting reliable negative examples
DDI	Drug-drug interactions
DILCA	Distance learning for categorical attributes
EKG	Electro-cardiogram
EM	Expectation maximization algorithm
EM-NB	Expectation maximization naïve Bayes algorithm
FBC	Full Bayesian network classifier
GAN	Generative adversarial network
GenPU	Generative adversarial positive-unlabeled learning
GLLC	Global and local learning algorithm
GMM	Gaussian mixture models
HNB	Hidden naïve Bayes
HSI	Hyperspectral imaging
IoT	Internet of things

Table B.1: (*Continued*)

Acronym	Meaning
kNN	k nearest neighbors
LDCE	Loss decomposition and centroid estimation
LGN	Learning by generating negative examples
LLSVM	Large-margin label-calibrated SVM
LOF	Local outlier factors
MCLS	Maximum margin clustering with least squares SVM
ML	Machine learning
MLE	Maximum likelihood estimation
MLR	Modified logistic regression
MMPU	Multi-manifold PU learning
MPE	Mixture proportion estimation
NB	Naïve Bayes algorithm
NLP	Natural language processing
OCC	One-class classification
OCSVM	One-class SVM
PAODE	PU averaged one-dependence estimators
PEBL	Positive example based learning
PE-PUC	Positive document enlarging PU classifier
PFBC	PU full Bayesian network classifier
PGAN	Positive-Gan
PGPU	Probabilistic-gap PU model
PHNB	PU hidden naïve Bayes
PNB algorithm	Positive naïve Bayes algorithm
PNLH	Positive examples and negative examples labeling heuristics
PR curve	Precision-recall curve
PSoL	Positive sample only learning
PTAN algorithm	Positive tree augmented naïve Bayes algorithm
PU Learning	Positive unlabeled learning

Table B.1: (*Continued*)

Acronym	Meaning
PUF-score	Positive unlabeled estimated F-score metric from Section 4.5
PURL	Positive-unlabeled reward learning
RGB	Red green blue
RNN	Recurrent neural network
ROC curve	Receiver operating characteristic curves
ROC-SVM	Rocchio-support vector machine
SAR	Selected at random
SAR-EM	Selected at random expectation maximization
SCAR	Selected completely at random
SCC	Single-class classification
SVM	Support vector machine
S-EM (or Spy-EM)	Spy expectation maximization
TFIDF	Term frequency-inverse documentation frequency
TFIPNDF	Term frequency-inverse positive-negative document frequency
TIcE	Tree induction for c estimation
uLSIF	Unconstrained least-squares importance fitting
UPTAN	Uncertain positive tree augmented naïve Bayes
USMO	Unlabeled data in sequential minimal optimization algorithm
WVC	Weighted voting classifier

APPENDIX C

Publicly Accessible Code Repositories

While most authors do not make their code publicly accessible, it is becoming more common to do so, typically through public GitHub repositories. The following papers had publicly available code repositories that were confirmed as of the time of this writing unless otherwise mentioned.

Table C.1

Paper	Code Repository
De Comité et al., 1999	ftp://grappa.univ-lille3.fr/pub/Experiments/C45PosUnl (not confirmed)
Bekker and Davis, 2019	https://github.com/ML-KULeuven/SAR-PU
Kato, Teshima, and Honda, 2019	https://github.com/MasaKat0/PUlearning
Li, Liu, and Ng, 2010	https://www.cs.uic.edu/~liub/LPU/LPU-download.html (executable only, no code)
Northcutt, Wu, and Chuang, 2017	https://github.com/cgnorthcutt/rankpruning
Kiryo et al., 2017	https://github.com/kiryor/nnPUlearning, https://github.com/GarrettLee/nnpu_tf
Liu et al., 2002	https://www2.cs.uic.edu/~liub/S-EM/readme.html (executable only, no code)
Kwon et al., 2020	https://github.com/eraser347/WMMD_PU
Claesen, De Smet, et al., 2015	https://github.com/claesenm/resvm
Bekker and Davis, 2018a	https://dtai.cs.kuleuven.be/software/tice
Ramaswamy, Scott, and Tewari, 2016	http://web.eecs.umich.edu/~cscott/code.html#kmpe
Sansone, De Natale, and Zhou, 2019	https://github.com/emsansone/USMO
Teisseyre, Mielniczuk, and Łazęcka, 2020	https://github.com/teisseyrep/PUlogistic
Yang, Liu, and Yang, 2017	https://github.com/PengyiYang/AdaSampling
Zuluaga et al., 2011	http://www.creatis.insa-lyon.fr/software/public/DLDalgorithms
Begum et al., 2013	https://www.cs.ucr.edu/~nbegu001/SSL_myMDL.htm

APPENDIX D

Selected Publicly Available Datasets Referenced in PU Papers

While a vast number of public and private datasets were used by the papers referenced in this book, due to space requirements we have included only those of each type that are most representative. For similar reasons, only the name of the datasets and the papers that used them are given here.

Table D.1: (*Continues*)

Simulated datasets	• Concentric circles: (Hou et al., 2018) • TreeData: (Qin et al., 2013) • TwoLines, TwoSpirals, TwoKnots, Roll&Plane: (Gong, Shi, et al., 2019) • TwoMoons: (Ke et al., 2018), (Gong, Shi, et al., 2019), (Kwon et al., 2020), (Hou et al., 2018)
General classification datasets	• Adult UCI: (De Comité et al., 1999), (Bekker and Davis, 2018a), (Bekker and Davis, 2018b), (Bekker and Davis, 2019), (Blanchard, Lee, and Scott, 2010), (Denis, Gilleron, and Letouzey, 2005), (Teisseyre, Mielniczuk, and Łazęcka, 2020) • Cover Type: (Bekker and Davis, 2019), (Claesen, De Smet, et al., 2015), (Bekker and Davis, 2018a), (Bekker and Davis, 2018b), (Ramola, Jain, and Radivojac, 2019) • Credit-a UCI: (Gan, Zhang, and Song, 2017), (Teisseyre, Mielniczuk, and Łazęcka, 2020), (Ienco and Pensa, 2016) • Mushrooms: (Kato et al., 2018), (Bekker and Davis, 2018a), (Ramola, Jain, and Radivojac, 2019), (Bekker and Davis, 2018b), (Shi et al., 2018), (Bekker and Davis, 2019), (Gan, Zhang, and Song, 2017), (Blanchard, Lee, and Scott, 2010), (Jain, White and Radivojac, 2017), (Ramola, Jain, and Radivojac, 2019), (Jain, White, and Radivojac, 2016), (Jain et al., 2016), (Ienco and Pensa, 2016) • Pima Indians Diabetes: (Jain, White, and Radivojac, 2017), (Ienco and Pensa, 2016), (Jain et al., 2016), (Jain, White, and Radivojac, 2016), (Yang, Liu, and Yang, 2017) • Vote UCI: (Shi et al., 2018), (Gan, Zhang, and Song, 2017), (Teisseyre, Mielniczuk, and Łazęcka, 2020), (Ienco and Pensa, 2016) • Wine: (Jain, White, and Radivojac, 2017), (Jain et al., 2016), (Jain, White, and Radivojac, 2016)

Table D.1: (*Continued*)

Anomaly datasets	• Epileptic Seizure Recognition UCI: (Ramola, Jain, and Radivojac, 2019)
Biomedical datasets	• Transporter Classification (TCDB) and SwissProt: (Elkan and Noto, 2008), (He et al., 2018) • RegulonDB Gene regulatory databases: (Mordelet and Vert, 2014), (Geurts, 2011), (Cerulo, Elkan, and Ceccarelli, 2010) • Breast cancer UCI dataset: (Bekker and Davis, 2019), (Bekker and Davis, 2018b), (Yu, 2005), (Gan, Zhang, and Song, 2017), (Shi et al., 2018), (Ienco and Pensa, 2016), (Bekker and Davis, 2018a), (Blanchard, Lee, and Scott, 2010), (Teisseyre, Mielniczuk, and Łazęcka, 2020), (Yang, Liu, and Yang, 2017) • Heart Disease Cleveland database: (Ke et al., 2018), (Gan, Zhang, and Song, 2017), (Ienco and Pensa, 2016), (Blanchard, Lee, and Scott, 2010), (Teisseyre, Mielniczuk, and Łazęcka, 2020)
Recommender systems	• Facebook-like Forum, MovieTweeting, Last.fm Music: (Chang et al., 2016)
Marketing datasets	• Bank UCI Dataset: (Jain, White, and Radivojac, 2017), (Jain et al., 2016), (Jain, White, and Radivojac, 2016)
Fraud/spam datasets	• Dianping Dataset: (Li et al., 2014) • Spambase: (Kato, Teshima, and Honda, 2019), (Jain, White, and Radivojac, 2017), (Kato et al., 2018), (Teisseyre, Mielniczuk, and Łazęcka, 2020), (Niu et al., 2016), (Jain et al., 2016), (Jain, White, and Radivojac, 2016), (Ienco and Pensa, 2016)
Image datasets	• MNIST: (Kato, Teshima, and Honda, 2019), (Zhang et al., 2019), (Northcutt, Wu, and Chuang, 2017), (Kato et al., 2018), (du Plessis, Niu, and Sugiyama, 2015), (Kiryo et al., 2017), (Jaskie, Elkan, and Spanias, 2019), (Gong, Shi, et al., 2019), (Hou et al., 2018), (Chiaroni et al., 2018), (Loghmani, Vincze, and Tommasi, 2020), (Claesen, De Smet, et al., 2015), (Zhang et al., 2017), (Xu et al., 2017) • USPS Handwritten Digits: (Kato, Teshima, and Honda, 2019), (Ke et al., 2018), (Shi et al., 2018), (Hou et al., 2018), (Loghmani, Vincze, and Tommasi, 2020), (Xu et al., 2017) • CIFAR-10: (Kato, Teshima, and Honda, 2019), (Kiryo et al., 2017), (Chiaroni et al., 2018), (Loghmani, Vincze, and Tommasi, 2020) • Statlog (Landsat Satellite) Dataset: (Zhang et al., 2019), (Jain, White, and Radivojac, 2017), (Jain et al., 2016), (Jain, White, and Radivojac, 2016)
Video datasets	• UMN Crowd Behavior Dataset: (Zhang et al., 2019), (Zhang et al., 2017) • HockeyFight Dataset: (Shi et al., 2018), (Gong, Shi, et al., 2019)

Table D.1: (*Continued*)

Remote sensing	• LANDSAT Thematic Mapper (TM): (Jeon and Landgrebe, 1999) • Jasper Ridge Hyperspectral Imaging: (Gong, Shi, et al., 2019)
Text datasets	• **20 Newsgroups:** (Liu et al., 2003), (Zhang and Lee, 2005), (Bekker and Davis, 2019), (Bekker and Davis, 2018b), (Li, Liu, and Ng, 2010), (Lee and Liu, 2003), (Nigam et al., 2000), (Mordelet and Vert, 2014), (Kiryo et al., 2017), (Han et al., 2016), (Li and Liu, 2005), (Bekker and Davis, 2018a), (Li et al., 2009) • **Reuters-21578:** (Liu et al., 2003), (Zhang and Lee, 2005), (Yu, 2005), (Li, Liu, and Ng, 2010), (Li and Liu, 2003), (Han et al., 2016), (Li and Liu, 2005) • **Pageblocks:** (Kato, Teshima, and Honda, 2019), (Jain, White, and Radivojac, 2017), (Jain et al., 2016), (Jain, White, and Radivojac, 2016)
Time series datasets	• **Sunspots:** (de Carvalho Pagliosa and de Mello, 2018) • **Statlog (Shuttle) dataset:** (Zhang et al., 2019), (Kato, Teshima, and Honda, 2019), (Jain, White, and Radivojac, 2017), (Jain et al., 2016), (Jain, White, and Radivojac, 2016) • **Waveforms UCI:** (Blanchard, Lee, and Scott, 2010), (Kato et al., 2018), (Niu et al., 2016)
Network intrusion datasets	• **Kyoto dataset:** (Zhang et al., 2019), (Zhang et al., 2017) • **KDDCUP'99:** (Qin et al., 2013)
Speech and audio	• **Phoneme:** (Gong, Liu, et al., 2019), (Niu et al., 20016)

APPENDIX E

Additional Resources

Table E.1

A PU Learning Tutorial by Jessa Bekker, author of (Bekker and Davis, 2018a, 2018b, 2019, 2020)	A six video tutorial series posted on January 3, 2021 at https://jessa.github.io/news/pul_tutorial/
A Positive Unlabeled Learning for TinyML Webinar by Kristen Jaskie, author of (Jaskie and Spanias, 2019; Jaskie, Elkan, and Spanias, 2019; Jaskie, Martin, and Spanias, 2021)	A video presentation to the TinyML community on March 9, 2021, posted on YouTube on March 11th at https://www.youtube.com/watch?v=uk6SlTzfbUY
Positive and Unlabelled Learning: Recovering Labels for Data Using Machine Learning A problem walk-through and Python implementation by Aaron Ward	https://heartbeat.fritz.ai/positive-and-unlabelled-learning-recovering-labels-for-data-using-machine-learning-59c1def5452f
Positive-unlabeled learning A problem walk-through and Python implementation by Roy Wright	https://roywrightme.wordpress.com/2017/11/16/positive-unlabeled-learning

Bibliography

Angluin, D. and Laird, P. (1988). Learning from noisy examples, *Machine Learning*, 2(4):343–370. DOI: 10.1023/A:1022873112823. 32

Arjannikov, T. (2021). Cold-start hospital length of stay prediction using positive-unlabeled learning, *IEEE EMBS International Conference on Biomedical and Health Informatics*, Athens, Greece. DOI: 10.1109/bhi50953.2021.9508596. 84

Arjannikov, T. and Tzanetakis, G. (2021). An empirical investigation of PU learning for predicting length of stay, *IEEE International Conference on Healthcare Informatics*, pages 41–47, Victoria, BC. DOI: 10.1109/ichi52183.2021.00019. 84

Austin, P. C. (2011). An introduction to propensity score methods for reducing the effects of confounding in observational studies, *Multivariate Behavioral Research*, 46(3):399–424. DOI: 10.1080/00273171.2011.568786. 24

Barcaccia, G., Lucchin, M., and Cassandro, M. (2016). DNA barcoding as a molecular tool to track down mislabeling and food piracy, *Diversity*, 8(1). DOI: 10.3390/d8010002. 32

Begum, N., Hu, B., Rakthanmanon, T., and Keogh, E. (2013). Towards a minimum description length based stopping criterion for semi-supervised time series classification, *Proc. of the IEEE 14th International Conference on Information Reuse and Integration*, pages 333–340, IEEE IRI. DOI: 10.1109/iri.2013.6642490. 84, 95, 105

Bekker, J. and Davis, J. (2018a). Estimating the class prior in positive and unlabeled data through decision tree induction, *The 32nd AAAI Conference on Artificial Intelligence (AAAI)*, pages 2712–2719, AAAI Press, New Orleans, LA. 75, 83, 84, 92, 105, 107, 108, 109, 111

Bekker, J. and Davis, J. (2018b). Learning from positive and unlabeled data under the selected at random assumption, *Proc. of the 2nd International Workshop on Learning with Imbalanced Domains: Theory and Applications*, pages 94:8–22, PMLR, Dublin, Ireland. 57, 58, 59, 84, 92, 107, 108, 109, 111

Bekker, J. and Davis, J. (2019). Beyond the selected completely at random assumption for learning from positive and unlabeled data, *Proc. of the European Conference on Machine Learning and Principles and Practice of Knowledge Discovery in Databases (ECML PKDD)*, 11907, pages 71–85. http://arxiv.org/abs/1809.03207 DOI: 10.1007/978-3-030-46147-8_5. 24, 26, 27, 37, 57, 58, 59, 84, 105, 107, 108, 109, 111

Bekker, J. and Davis, J. (2020). Learning from positive and unlabeled data: A survey, *Machine Learning*, 109(4):719–760. DOI: 10.1007/s10994-020-05877-5. 21, 24, 26, 45, 49, 61, 75, 111

Bepler, T., Morin, A., Rapp, M., et al. (2019). Positive-unlabeled convolutional neural networks for particle picking in cryo-electron micrographs, *Nature Methods*, 16(11):1153–1160. DOI: 10.1038/s41592-019-0575-8. 84

Birkenes, Ø., Matsui, T., Tanabe, K., and Myrvoll, T. A. (2007). N-best rescoring for speech recognition using penalized logistic regression machines with garbage class, *ICASSP*, 4:449–452. DOI: 10.1109/icassp.2007.366946. 94

Bishop, C. (2011). *Pattern Recognition and Machine Learning, Information Science and Statistics*, Springer, New York. 97

Blanchard, G., Lee, G., and Scott, C. (2010). Semi-supervised novelty detection, *Journal of Machine Learning Research*, 11:2973–3009. 81, 107, 108, 109

Blum, A. and Mitchell, T. (1998). Combining labeled and unlabeled data with co-training, *Proc. of the 11th Annual Conference on Computational Learning Theory*, pages 92–100. DOI: 10.1145/279943.279962. 92

Breiman, L. (1996). Bagging predictors, *Machine Learning*, 8:123–140. DOI: 10.1007/bf00058655. 72

Buzo, A., Gray, A. H., Gray, R. M., and Markel, J. D. (1979). A two-step speech compression system with vector quantizing, *ICASSP*, 4:52–55. DOI: 10.1109/icassp.1979.1170755. 94

Calvo, B. (2008). Positive unlabelled learning with applications in computational biology, *Department of Computer Science and Artificial Intelligence*, University of the Basque Country. 74, 82

Calvo, B., Larrañaga, P., and Lozano, J. A. (2007). Learning Bayesian classifiers from positive and unlabeled examples, *Pattern Recognition Letters*, 28(16):2375–2384. DOI: 10.1016/j.patrec.2007.08.003. 49

de Campos, L. M., Fernández-Luna, J. M., Huete, J. F., and Redondo-Expósito, L. (2018). Positive unlabeled learning for building recommender systems in a parliamentary setting, *Information Sciences*, 433–434:221–32. DOI: 10.1016/j.ins.2017.12.046. 87

de Carvalho Pagliosa, L. and de Mello, R. F. (2018). Semi-supervised time series classification on positive and unlabeled problems using cross-recurrence quantification analysis, *Pattern Recognition*, 80:53–63. DOI: 10.1016/j.patcog.2018.02.030. 95, 109

Cerulo, L., Elkan, C., and Ceccarelli, M. (2010). Learning gene regulatory networks from only positive and unlabeled data, *BMC Bioinformatics*, 11:228. DOI: 10.1186/1471-2105-11-228. 83, 107

Chang, S., Zhang, Y., Tang, J., et al. (2016). Positive-unlabeled learning in streaming networks, *KDD*, pages 755–764, ACM, San Francisco, CA. DOI: 10.1145/2939672.2939744. 87, 88, 92, 95, 108

Chapelle, O., Schölkopf, B., and Zien, A. (2006). *Semi-Supervised Learning*. Edited by O. Chapelle, B. Schölkopf, and A. Zien. MIT Press. DOI: 10.7551/mit-press/9780262033589.001.0001. 5, 29, 98

Chaudhari, S. and Shevade, S. (2012). Learning from positive and unlabelled examples using maximum margin clustering, *Lecture Notes in Computer Science (including subseries Lecture Notes in Artificial Intelligence and Lecture Notes in Bioinformatics), 7665 LNCS (PART 3)*, pages 465–473. DOI: 10.1007/978-3-642-34487-9_56. 64, 66

Chiaroni, F., Rahal, M. C., Hueber, N., and Dufaux, F. (2018). Learning with a generative adversarial network from a positive unlabeled dataset for image classification, *25th IEEE International Conference on Image Processing (ICIP)*, pages 1368–1372, Athens, Greece. DOI: 10.1109/icip.2018.8451831. 70, 71, 89, 108

Chine, W., Mellit, A., Lughi, V., et al. (2016). A novel fault diagnosis technique for photovoltaic systems based on artificial neural networks, *Renewable Energy*, 90:501–512. DOI: 10.1016/j.renene.2016.01.036. 93

Claesen, M., De Smet, F., Suykens, J. A. K., and De Moor, B. (2015). A robust ensemble approach to learn from positive and unlabeled data using SVM base models, *Neurocomputing*, 160:73–84. DOI: 10.1016/j.neucom.2014.10.081. 73, 82, 85, 92, 105, 107, 108

Claesen, M., Davis, J., De Smet, F., and De Moor, B. (2015). Assessing binary classifiers using only positive and unlabeled data, pages 1–14. http://arxiv.org/abs/1504.06837 45

De Comité, F., Denis, F., Gilleron, R., and Letouzey, F. (1999). Positive and unlabeled examples help learning, *Conference on Algorithic Learning Theory*, 1720(December):219–230. DOI: 10.1007/3-540-46769-6_18. 5, 23, 47, 74, 84, 92, 105, 107

Crowe (2019). Fraud costs the global economy over US$5 trillion, Crowe.com. https://www.crowe.com/global/news/fraud-costs-the-global-economy-over-us\protect\TU\textdollar5-trillion (Accessed: June 5, 2021). 88

Das, S., Saier, M. H., and Elkan, C. (2007). Finding transport proteins in a general protein database, *Proc. of the 11th European Conference on Principles and Practice of Knowledge Discovery in Databases (ECML PKDD)*, pages 54–66, Springer, Warsaw, Poland. DOI: 10.1007/978-3-540-74976-9_9. 83

Deng, L., Seltzer, M., Yu, D., et al. (2010). Binary coding of speech spectrograms using a deep auto-encoder, *Proc. of the 11th Annual Conference of the International Speech Communication Association, INTERSPEECH*, pages 1692–1695. DOI: 10.21437/interspeech.2010-487. 94

Deng, X., Li, W., Liu, X., Guo, Q., and Newsam, S. (2018). One-class remote sensing classification: One-class vs. binary classifiers, *International Journal of Remote Sensing*, 39(6):1890–1910. DOI: 10.1080/01431161.2017.1416697. 28, 31

Denis, F. (1998). PAC learning from positive statistical queries, *Lecture Notes in Computer Science (including subseries Lecture Notes in Artificial Intelligence and Lecture Notes in Bioinformatics)*, 1501:112–126. DOI: 10.1007/3-540-49730-7_9. 47, 65, 84, 87

Denis, F., Laurent, A., Gilleron, R., and Tommasi, M. (2003). Text classification and co-training from positive and unlabeled examples, *The International Conference on Machine Learning—Workshop: The Continuum from Labeled to Unlabeled Data (ICML)*, pages 80–87, Washington, DC. 74, 92

Denis, F., Gilleron, R., and Letouzey, F. (2005). Learning from positive and unlabeled examples, *Theoretical Computer Science*, 348(1):70–83. DOI: 10.1016/j.tcs.2005.09.007. 84, 107

Denis, F., Gilleron, R., and Tommasi, M. (2002). Text classification from positive and unlabeled examples, *The 9th International Conference on Information Processing and Management of Uncertainty in Knowledge-Based (IPMU)*, Annecy, France. 49, 74, 92

Diabetes (2018). World Health Organization. https://www.who.int/news-room/fact-sheets/detail/diabetes 13

Domingues, I., Amorim, J. P., Abreu, P. H., Duarte, H., and Santos, J. (2018). Evaluation of oversampling data balancing techniques in the context of ordinal classification, *Proc. of the International Joint Conference on Neural Networks*. DOI: 10.1109/ijcnn.2018.8489599. 39

Elkan, C. (2001). The foundations of cost-sensitive learning, *Proc. of the 17th International Joint Conference on Artificial Intelligence*, pages 973–978, Seattle, WA. 39, 50, 51

Elkan, C. and Noto, K. (2008). Learning classifiers from only positive and unlabeled data, *Proc. of the 14th International Conference on Knowledge Discovery and Data Mining (SIGKDD)*, pages 213–20, ACM, Las Vegas, NV. DOI: 10.1145/1401890.1401920. 22, 23, 51, 53, 54, 74, 83, 107

Fan, J., Rao, S., Muniraju, G., Tepedelenlioglu, C., and Spanias, A. (2020). Fault classification in photovoltaic arrays via graph signal processing, *ICPS*, IEEE, Tampere, Finland. 93

Frénay, B. and Verleysen, M. (2014). Classification in the presence of label noise: A survey, *IEEE Transactions on Neural Networks and Learning Systems*, 25(5):845–869. DOI: 10.1109/tnnls.2013.2292894. 32, 33, 50

Fung, G. P. C., Yu, J. X., Lu, H., and Yu, P. S. (2006). Text classification without negative examples revisit, *IEEE Transactions on Knowledge and Data Engineering (TKDE)*, 18(1):6–20. DOI: 10.1109/tkde.2006.16. 61, 63, 67

Gan, H., Zhang, Y., and Song, Q. (2017). Bayesian belief network for positive unlabeled learning with uncertainty, *Pattern Recognition Letters*, 90:28–35. DOI: 10.1016/j.patrec.2017.03.007. 50, 86, 107, 108

Garg, P. and Sundararajan, S. (2009). Active learning in partially supervised classification, *International Conference on Information and Knowledge Management, Proc.*, pages 1783–1786. DOI: 10.1145/1645953.1646229. 74

Garoudja, E., Chouder, A., Kara, K., and Silvestre, S. (2017). An enhanced machine learning based approach for failures detection and diagnosis of PV systems, *Energy Conversion and Management*, 151(September):496–513. DOI: 10.1016/j.enconman.2017.09.019. 93

Geiss, L. S., Bullard, K. M. K., Brinks, R., Hoyer, A., and Gregg, E. W. (2018). Trends in type 2 diabetes detection among adults in the USA, 1999–2014, *BMJ Open Diabetes Research and Care*, 6(1):1–5. DOI: 10.1136/bmjdrc-2017-000487. 13

Geurts, P. (2011). Learning from positive and unlabeled examples by enforcing statistical significance, *Journal of Machine Learning Research*, 15:305–314. 82, 83, 107

Gomez-Uribe, C. A. and Hunt, N. (2015). The netflix recommender system: Algorithms, business value, and innovation, *ACM Transactions on Management Information Systems*, 6(4). DOI: 10.1145/2843948. 87

Gong, C., Liu, T., Yang, J., and Tao, D. (2019). Large-margin label-calibrated support vector machines for positive and unlabeled learning, *IEEE Transactions on Neural Networks and Learning Systems*, 30(11):3471–3483. DOI: 10.1109/tnnls.2019.2892403. 70, 88, 91, 92, 109

Gong, C., Shi, H., Yang, J., Yang, J., and Yanga, J. (2019). Multi-manifold positive and unlabeled learning for visual analysis, *IEEE Transactions on Circuits and Systems for Video Technology (TCSVT)*, pages 1–14. DOI: 10.1109/tcsvt.2019.2903563. 37, 65, 67, 70, 89, 90, 91, 92, 107, 108

González, M., Bergmeir, C., Triguero, I., Rodríguez, Y., and Benítez, J. M. (2016). On the stopping criteria for k-Nearest neighbor in positive unlabeled time series classification problems, *Information Sciences*, 328:42–59. DOI: 10.1016/j.ins.2015.07.061. 95

Goodfellow, I., Bengio, Y., and Courville, A. (2016). *Deep Learning*, The MIT Press (Adaptive Computation and Machine Learning Series). 98

Gray, R. M. (1984). Vector quantization, *IEEE ASSP Magazine*, pages 4–29. DOI: 10.1109/MASSP.1984.1162229. 94

Guo, Q., Li, W., Liu, D., and Chen, J. (2012). A framework for supervised image classification with incomplete training samples, *Photogrammetric Engineering and Remote Sensing*, 78(6):595–604. DOI: 10.14358/pers.78.6.595. 28, 29, 31

Hameed, P. N., Verspoor, K., Kusljic, S., and Halgamuge, S. (2017). Positive-unlabeled learning for inferring drug interactions based on heterogeneous attributes, *BMC Bioinformatics*, 18(1):1–15. DOI: 10.1186/s12859-017-1546-7. 85

Han, J., Zuo, W., Liu, L., Xu, Y., and Peng, T. (2016). Building text classifiers using positive, unlabeled and "outdated" examples, *Concurrency and Computation: Practice and Experience*, 28(13):3691–3706. DOI: 10.1002/cpe.3879. 68, 73, 92, 109

Hastie, T., Tibshirani, R., and Friedman, J. (2016). *The Elements of Statistical Learning: Data Mining, Inference, and Prediction*, 2nd ed., Springer (Series in Statistics). 98

He, F., Webb, G. I., Liu, T., and Tao, D. (2018). Instance-dependent PU learning by Bayesian optimal relabeling, *ArXiv* [Preprint]. https://arxiv.org/abs/1808.02180v1 27, 50, 61, 62, 65, 83, 107

He, J., Zhang, Y., Li, X., and Wang, Y. (2011). Bayesian classifiers for positive unlabeled learning, *Lecture Notes in Computer Science (including subseries Lecture Notes in Artificial Intelligence and Lecture Notes in Bioinformatics)*, 6897 LNCS(60873196):81–93. DOI: 10.1007/978-3-642-23535-1_9. 49

Hernández-González, J., Inza, I., and Lozano, J. A. (2017). Learning from proportions of positive and unlabeled examples, *International Journal of Intelligent Systems*, 32(2):109–33. DOI: 10.1002/int.21832. 86

Hoffmann, F., Bertram, T., Mikut, R., Reischl, M., and Nelles, O. (2019). Benchmarking in classification and regression, *Wiley Interdisciplinary Reviews: Data Mining and Knowledge Discovery*, 9(5):1–17. DOI: 10.1002/widm.1318. 36

Hou, M., Chaib-Draa, B., Li, C., and Zhao, Q. (2018). Generative adversarial positive-unlabelled learning, *Proc. of the 27th International Joint Conference on Artificial Intelligence (IJCAI)*, Stockholm, Sweden. DOI: 10.24963/ijcai.2018/312. 37, 70, 71, 81, 91, 107, 108

Hu, W., Le, R., Liu, B., et al. (2021). Predictive adversarial learning from positive and unlabeled data, *Proc. of the AAI Conference on Artificial Intelligence*, 35(9):7806–7814. 70, 71

Ienco, D. and Pensa, R. G. (2016). Positive and unlabeled learning in categorical data, *Neurocomputing*, 196:113–124. DOI: 10.1016/j.neucom.2016.01.089. 64, 67, 84, 92, 107, 108

Ienco, D., Pensa, R. G., and Meo, R. (2012). From context to distance: Learning dissimilarity for categorical data clustering, *ACM Transactions on Knowledge Discovery from Data*, 6(1). DOI: 10.1145/2133360.2133361. 64, 67

Imbens, G. W. and Rubin, D. B. (2015). *Causal Inference for Statistics, Social, and Biomedical Sciences*, Cambridge University Press. DOI: 10.1017/cbo9781139025751. 24

Jain, S., White, M., Trosset, M. W., and Radivojac, P. (2016). Nonparametric semi-supervised learning of class proportions, *ArXiv Preprint ArXiv:1601.01944*. 77, 83, 87, 107, 108, 109

Jain, S., White, M., and Radivojac, P. (2016). Estimating the class prior and posterior from noisy positives and unlabeled data, *Advances in Neural Information Processing Systems*, 29:2693–2701. 77, 83, 87, 107, 108, 109

Jain, S., White, M., and Radivojac, P. (2017). Recovering true classifier performance in positive-unlabeled learning, *31st AAAI Conference on Artificial Intelligence*, pages 2066–2072. 43, 44, 45, 83, 87, 107, 108, 109

Jannach, D. and Jugovac, M. (2019). Measuring the business value of recommender systems, *ACM Transactions on Management Information Systems*, 10(4):1–22. DOI: 10.1145/3370082. 87, 88

Japkowicz, N. and Stephen, S. (1998). The class imbalance problem: A systematic study, *Drugs and Therapy Perspectives*, 12(7):10. DOI: 10.3233/ida-2002-6504. 39

Jaskie, C. and Fette, B. (1992). A survey of low bit rate vocoders, *Proc. of Voice Systems Worldwide*, pages 35–46, London. 94

Jaskie, K., Elkan, C., and Spanias, A. (2019). A modified logistic regression for positive and unlabeled learning, *IEEE Asilomar*, pages 0–5, Pacific Grove, CA. DOI: 10.1109/ieeeconf44664.2019.9048765. 53, 55, 56, 57, 75, 82, 88, 89, 90, 92, 94, 108, 111

Jaskie, K., Martin, J., and Spanias, A. (2021). Photovoltaic fault detection using positive unlabeled learning, *Applied Sciences, MDPI AG*, 11(12):5599. DOI: 10.3390/app11125599. xv, 93, 111

Jaskie, K. and Spanias, A. (2019). Positive and unlabeled learning algorithms and applications: A survey, *IEEE IISA*, pages 1–8, Patras, Greece. DOI: 10.1109/iisa.2019.8900698. 87, 88, 91, 111

Jeon, B. and Landgrebe, D. A. (1999). Partially supervised classification using weighted unsupervised clustering, *IEEE Transactions on Geoscience and Remote Sensing*, 37(2 II):1073–1079. DOI: 10.1109/36.752225. 28, 29, 31, 64, 67, 91, 93, 108

Jiang, C., Yu, H.-F., Hsieh, C.-J., and Chang, K.-W. (2018). Learning word embeddings for low-resource languages by PU learning, *ACL Anthology*, pages 1024–1034. DOI: 10.18653/v1/n18-1093. 94

Joachims, T. (1997). A probabilistic analysis of the Rocchio algorithm with TFIDF for text categorization, *International Conference on Machine Learning (ICML)*, pages 143–151. 47

Kaboutari, A., Bagherzadeh, J., and Kheradmand, F. (2014). An evaluation of two-step techniques for positive-unlabeled learning in text classification, *International Journal of Computer Applications Technology and Research*, 3(9):592–594. DOI: 10.7753/ijcatr0309.1012. 61

Kanehira, A. and Harada, T. (2016). Multi-label ranking from positive and unlabeled data, *Proc. of the IEEE Computer Society Conference on Computer Vision and Pattern Recognition*, pages 5138–5146. DOI: 10.1109/cvpr.2016.555. 77, 89

Kato, M., Xu, L., Niu, G., and Sugiyama, M. (2018). Alternate estimation of a classifier and the class-prior from positive and unlabeled data, *ArXiv*, pages 1–19. 52, 78, 93, 107, 108, 109

Kato, M., Teshima, T., and Honda, J. (2019). Learning from positive and unlabeled data with a selection bias, *The International Conference on Learning Representations*, ICLR, New Orleans, LA. 19, 27, 28, 37, 59, 60, 61, 81, 89, 93, 105, 108, 109

Ke, T., Jing, L., Lv, H., Zhang, L., and Hu, Y. (2018). Global and local learning from positive and unlabeled examples, *Applied Intelligence*, 48(8):2373–2392. DOI: 10.1007/s10489-017-1076-z. 37, 50, 70, 83, 89, 92, 107, 108

Kearns, M. and Li, M. (1988). Learning in the presence of malicious errors, *Proc. of the Annual ACM Symposium on Theory of Computing*, pages 267–280. DOI: 10.1145/62212.62238. 32, 33

Khan, S. S. and Madden, M. G. (2014). One-class classification: Taxonomy of study and review of techniques, *Knowledge Engineering Review*, pages 345–374, Cambridge University Press. DOI: 10.1017/s026988891300043x. 28, 31

Kiryo, R., Niu, G., du Plessis, M. C., and Sugiyama, M. (2017). Positive-unlabeled learning with non-negative risk estimator, *31st International Conference on Neural Information Processing Systems (NIPS)*, pages 1674–84, Long Beach, CA, Curran Assoc. Inc. http://arxiv.org/abs/1703.00593 52, 53, 59, 60, 70, 71, 105, 108, 109

Kumar, A. and Raj, B. (2016). Audio event detection using weakly labeled data, *MM—Proc. of the ACM Multimedia Conference*, pages 1038–1047, ACM Press, New York. DOI: 10.1145/2964284.2964310. 94

Kwon, Y., Kim, W., Sugiyama, M., and Paik, M. C. (2020). Principled analytic classifier for positive-unlabeled learning via weighted integral probability metric, *Machine Learning*, 109(3):513–532. DOI: 10.1007/s10994-019-05836-9. 37, 70, 81, 82, 89, 92, 105, 107

Lappas, T. (2012). Fake reviews: The malicious perspective, *NLDB*, pages 23–34. DOI: 10.1007/978-3-642-31178-9_3. 32

Lee, W. S. and Liu, B. (2003). Learning with positive and unlabeled examples using weighted logistic regression, *Proc. of the 12th International Conference on Machine Learning (ICML)*, pages 448–455, AAAI Press, Washington, DC. 33, 42, 43, 50, 87, 92, 109

Li, B., Delpha, C., Diallo, D., and Migan-Dubois, A. (2021). Application of artificial neural networks to photovoltaic fault detection and diagnosis: A review, *Renewable and Sustainable Energy Reviews*, 138(October). DOI: 10.1016/j.rser.2020.110512. 93

Li, H., Chen, Z., Liu, B., Wei, X., and Shao, J. (2014). Spotting fake reviews via collective positive-unlabeled learning, *Proc. of the 14th International Conference on Data Mining (ICDM)*, pages 899–904, IEEE, Washington, DC. DOI: 10.1109/icdm.2014.47. 88, 108

Li, M., Pan, S., Zhang, Y., and Cai, X. (2016). Classifying networked text data with positive and unlabeled examples, *Pattern Recognition Letters*, 77:1–7. DOI: 10.1016/j.patrec.2016.03.006. 50, 92

Li, W., Guo, Q., and Elkan, C. (2011a). A positive and unlabeled learning algorithm for one-class classification of remote-sensing data, *IEEE Transactions on Geoscience and Remote Sensing*, 49(2):717–725. DOI: 10.1109/tgrs.2010.2058578. 29, 31, 86, 89, 91

Li, W., Guo, Q., and Elkan, C. (2011b). Can we model the probability of presence of species without absence data?, *Ecography*, 34(6):1096–1105. DOI: 10.1111/j.1600-0587.2011.06888.x. 85

Li, W., Guo, Q., and Elkan, C. (2020). One-class remote sensing classification from positive and unlabeled background data, *IEEE Journal of Selected Topics in Applied Earth Observations and Remote Sensing*, page 1. DOI: 10.1109/jstars.2020.3025451. 29, 30, 91

Li, X.-L., Yu, P. S., Liu, B., and Ng, S.-K. (2009). Positive unlabeled learning for data stream classification, *SIAM International Conference on Data Mining (SDM)*, pages 259–270, Sparks, SIAM. DOI: 10.1137/1.9781611972795.23. 95, 109

Li, X.-L. and Liu, B. (2005). Learning from positive and unlabeled examples with different data distributions, *European Conference on Machine Learning and Principles and Practice of Knowledge Discovery in Databases (ECML PKDD)*, pages 218–229. DOI: 10.1007/11564096_24. 69, 92, 109

Li, X. and Liu, B. (2003). Learning to classify texts using positive and unlabeled data, *The 18th International Joint Conference on Artificial Intelligence (IJCAI)*, pages 587–592, Morgan Kaufmann, Acapulco, Mexico. 61, 63, 66, 68, 92, 109

Li, X. L., Zhang, L., Liu, B., and Ng, S. K. (2010). Distributional similarity vs. PU learning for entity set expansion, *ACL–48th Annual Meeting of the Association for Computational Linguistics, Proc. of the Conference*, pages 359–364. 94

Li, X. L., Liu, B., and Ng, S. K. (2007). Learning to identify unexpected instances in the test set, *IJCAI International Joint Conference on Artificial Intelligence*, pages 2802–2807. 61, 63, 67

Li, X. L., Liu, B., and Ng, S. K. (2010). Negative training data can be harmful to text classification, *Proc. of the Conference on Empirical Methods in Natural Language Processing*, pages 218–228. 29, 92, 105, 109

Liang, C., Zhang, Y., Shi, P., and Hu, Z. (2012). Learning very fast decision tree from uncertain data streams with positive and unlabeled samples, *Information Sciences*, 213:50–67. DOI: 10.1016/j.ins.2012.05.023. 95

Ling, C. X. and Sheng, V. S. (2008). Cost-sensitive learning and the class imbalance problem, *Encyclopedia of Machine Learning*, pages 231–235. 39

Lippmann, R. P. (1989). Review of neural networks for speech recognition, *Neural Computation*, 1(1):1–38. DOI: 10.1162/neco.1989.1.1.1. 94

Liu, B., Lee, W. S., Yu, P. S., and Li, X. (2002). Partially supervised classification of text documents, *Proc. 19th International Conference on Machine Learning (ICML)*, pages 387–394, Morgan Kaufmann, Sydney, Australia. 61, 62, 66, 67, 92, 105

Liu, B., Dai, Y., Li, X., Lee, W. S., and Yu, P. S. (2003). Building text classifiers using positive and unlabeled examples, *Proc. of the 3rd IEEE International Conference on Data Mining (ICDM)*, pages 179–186, Melbourne, FL. DOI: 10.1109/icdm.2003.1250918. 23, 33, 50, 92, 109

Liu, L. and Peng, T. (2014). Clustering-based method for positive and unlabeled text categorization enhanced by improved TFIDF, *Journal of Information Science and Engineering*, 30(5):1463–1481. DOI: 10.6688/JISE.2014.30.5.10. 61, 64, 66, 73

Liu, R., Li, W., Liu, X., et al. (2018). An ensemble of classifiers based on positive and unlabeled data in one-class remote sensing classification, *IEEE Journal of Selected Topics in Applied Earth Observations and Remote Sensing*, 11(2):572–584. DOI: 10.1109/jstars.2017.2789213. 28, 29, 31, 91

Liu, T. and Tao, D. (2016). Classification with noisy labels by importance reweighting, *IEEE Transactions on Pattern Analysis and Machine Intelligence (TPAMI)*, 38(3):447–461. DOI: 10.1109/tpami.2015.2456899. 50

Liu, Y., Qiu, S., Zhang, P., et al. (2017). Computational drug discovery with dyadic positive-unlabeled learning, *Proc. of the 17th SIAM International Conference on Data Mining, SDM*, pages 45–53. DOI: 10.1137/1.9781611974973.6. 85

Liu, Z., Shi, W., Li, D., and Qin, Q. (2006). Partially supervised classification: Based on weighted unlabeled samples support vector machine, *International Journal of Data Warehousing and Mining (IJDWM)*, 2(3):42–56. DOI: 10.4018/jdwm.2006070104. 91

Loghmani, M. R., Vincze, M., and Tommasi, T. (2020). Positive-unlabeled learning for open set domain adaptation, *Pattern Recognition Letters*, 136:198–204. DOI: 10.1016/j.patrec.2020.06.003. 72, 108

Makhoul, J., Roucos, S., and Gish, H. (1985). Vector quantization in speech coding, *Proc. of the IEEE*, 73(11):1551–1588. DOI: 10.1109/proc.1985.13340. 94

Marussy, K. and Buza, K. (2013). SUCCESS: A new approach for semi-supervised classification of time-series, *Lecture Notes in Computer Science (including subseries Lecture Notes in Artificial Intelligence and Lecture Notes in Bioinformatics), 7894 LNAI(PART 1)*, pages 437–447. DOI: 10.1007/978-3-642-38658-9_39. 95

Matthews, B. W. (1975). Comparison of the predicted and observed secondary structure of T4 phage lysozyme, *BBA–Protein Structure*, 405(2):442–451. DOI: 10.1016/0005-2795(75)90109-9. 39

Menon, A. K., Rooyen, B. Van, Oong, C. S., and Williamson, R. C. (2015). Learning from corrupted binary labels via class-probability estimation, *Proc. of the 32nd International Conference on Machine Learning (ICML)*, pages 125–34, JMLR, Lille, France. 33, 50

Metzen, J. H. (2015). Probability calibration, *Github*. https://jmetzen.github.io/2015-04-14/calibration.html 55

Mitchell, T. (1997). Machine learning, *International McGraw-Hill Higher Education (Computer Science)*. 98

Mordelet, F. and Vert, J. P. (2011). ProDiGe: Prioritization of disease genes with multitask machine learning from positive and unlabeled examples, *BMC Bioinformatics*, pages 1–15. DOI: 10.1186/1471-2105-12-389. 73, 82

Mordelet, F. and Vert, J. P. (2014). A bagging SVM to learn from positive and unlabeled examples, *Pattern Recognition Letters*, 36:201–209. DOI: 10.1016/j.patrec.2013.06.010. 49, 72, 73, 82, 93, 107, 109

Moreno, P. J. and Agarwal, S. (2003). An experimental study of semi-supervised EM algorithms in audio classification and speaker identification, *Proc. of ICML*, pages 1–10. 94

Moya, M. M. and Hush, D. R. (1996). Network constraints and multi-objective optimization for one-class classification, *Neural Networks*, 9(3):463–474. DOI: 10.1016/0893-6080(95)00120-4. 28

Muñoz-Marí, J., Bovolo, F., Gómez-Chova, L., Bruzzone, L., and Camp-Valls, G. (2010). Semisupervised one-class support vector machines for classification of remote sensing data, *IEEE Transactions on Geoscience and Remote Sensing*, 48(8):3188–3197. DOI: 10.1109/tgrs.2010.2045764. 28, 29

Nan, X., Bao, L., Zhao, Xiaosa, et al. (2017). EPuL: An enhanced positive-unlabeled learning algorithm for the prediction of pupylation sites, *Molecules*, 22(9). DOI: 10.3390/molecules22091463. 83

Narayanaswamy, V., Ayyanar, R., Spanias, A., Tepedelenlioglu, C., and Srinivasan, D. (2019). Connection topology optimization in photovoltaic arrays using neural networks, *International Conference on Industrial Cyber-Physical Systems (ICPS)*, pages 167–172, IEEE, Taipei, Taiwan. DOI: 10.1109/icphys.2019.8780242. 93

Natarajan, N., Dhillon, I. S., Ravikumar, P., and Tewari, A. (2013). Learning with noisy labels, *Advances in Neural Information Processing Systems*, pages 1–9. 32, 33, 50

National Institute of Diabetes and Digestive and Kidney Diseases (2016). *Pima Indians Diabetes Dataset, Kaggle*. https://www.kaggle.com/uciml/pima-indians-diabetes-database 13, 14

Nettleton, D. F., Orriols-Puig, A., and Fornells, A. (2010). A study of the effect of different types of noise on the precision of supervised learning techniques, *Artificial Intelligence Review*, 33(4):275–306. DOI: 10.1007/s10462-010-9156-z. 33

Nguyen, M. N., Li, X. L., and Ng, S. K. (2011). Positive unlabeled learning for time series classification, *The International Joint Conference on Artificial Intelligence (IJCAI)*, pages 1421–1426. DOI: 10.5591/978-1-57735-516-8/IJCAI11-240. 95

Niculescu-Mizil, A. and Caruana, R. (2005). Predicting good probabilities with supervised learning, *ICML—Proc. of the 22nd International Conference on Machine Learning*, (1999):625–632. DOI: 10.1145/1102351.1102430. 55

Nigam, K., McCallum, A. K., Thrun, S., and Mitchell, T. (2000). Text classification from labeled and unlabeled documents using EM, *Machine Learning*, (39):103–134. DOI: 10.1023/A:1007692713085. 47, 92, 109

Niu, G., Plessis, M. C. du, Sakai, T., Ma, Y., and Sugiyama, M. (2016). Theoretical comparisons of positive-unlabeled learning against positive-negative learning, *30th International Conference on Neural Information Processing Systems (NIPS)*, pages 1207–15, ACM, Barcelona, Spain. 47, 108, 109

Northcutt, C. G., Jiang, L., and Chuang, I. L. (2021). Confident learning: Estimating uncertainty in dataset labels, *Journal of Artificial Intelligence Research*, 70:1373–1411. DOI: 10.1613/jair.1.12125. 33

Northcutt, C. G., Wu, T., and Chuang, I. L. (2017). Learning with confident examples: Rank pruning for robust classification with noisy labels, *Proc. of the 33rd Conference Uncertainty in Artificial Intelligence*, Sydney, Australia. 33, 89, 105, 108

Peng, T., Zuo, W., and He, F. (2008). SVM based adaptive learning method for text classification from positive and unlabeled documents, *Knowledge and Information Systems*, 16(3):281–301. DOI: 10.1007/s10115-007-0107-1. 61, 63, 66, 73

Perera, K. S., Aung, Z., and Woon, W. L. (2014). Machine learning techniques for supporting renewable energy generation and integration: A survey, *International Workshop on Data Analytics for Renewable Energy Integration*, pages 81–96, Springer. DOI: 10.1007/978-3-319-13290-7_7. 93

Pham, A. T. and Raich, R. (2018). Differential privacy for positive and unlabeled learning with known class priors, *IEEE Statistical Signal Processing Workshop, SSP*, pages 658–662. DOI: 10.1109/ssp.2018.8450839. 83, 84, 88, 93

Platt, J. (1999). Probabilistic outputs for support vector machines and comparisons to regularized likelihood methods, *Advances in Large Margin Classifiers*, 10(3):61–74. 9, 55

du Plessis, M. C., Niu, G., and Sugiyama, M. (2014). Analysis of learning from positive and unlabeled data, *Proc. of the 27th International Conference on Neural Information Processing Systems*, pages 703–711, ACM, Montreal, Canada. 51, 52, 53, 59, 81

du Plessis, M. C., Niu, G., and Sugiyama, M. (2015). Convex formulation for learning from positive and unlabeled data, *The 7th Asian Conference on Machine Learning (ACML)*. pages 221–236, Springer, Hong Kong, China. 52, 53, 70, 89, 91, 108

du Plessis, M. C., Niu, G., and Sugiyama, M. (2015a). Class-prior estimation for learning from positive and unlabeled data, *Asian Conference on Machine Learning (ACML)*. pages 221–236, Hong Kong, China. DOI: 10.1007/s10994-016-5604-6. 77, 78

du Plessis, M. C., Niu, G., and Sugiyama, M. (2015b). Convex formulation for learning from positive and unlabeled data, *32nd International Conference on Machine Learning, ICML*, 2:1386–1394. 59

du Plessis, M. C. and Sugiyama, M. (2014a). Class prior estimation from positive and unlabeled data, *IEICE Transactions on Information and Systems*, E96-D(5):1358–1362. DOI: 10.1587/transinf.e97.d.1358. 77

du Plessis, M. C. and Sugiyama, M. (2014b). Semi-supervised learning of class balance under class-prior change by distribution matching, *Neural Networks*, 50:110–119. DOI: 10.1016/j.neunet.2013.11.010. 77

Qin, X., Zhang, Y., Li, C., and Li, X. (2013). Learning from data streams with only positive and unlabeled data, *Journal of Intelligent Information Systems*, 40(3):405–430. DOI: 10.1007/s10844-012-0231-6. 94, 95, 107, 109

Rabiner, L. R. (1989). A tutorial on hidden Markov models and selected applications in speech recognition, *Proc. of the IEEE*, 77(2):257–286. DOI: 10.1109/5.18626. 94

Ramaswamy, H. G., Scott, C., and Tewari, A. (2016). Mixture proportion estimation via Kernel embedding of distributions, *Proc. of the 33rd International Conference on Machine Learning (ICML)*, pages 2052–60, JMLR, New York. http://arxiv.org/abs/1603.02501 78, 105

Ramola, R., Jain, S., and Radivojac, P. (2019). Estimating classification accuracy in positive-unlabeled learning: Characterization and correction strategies, *Pacific Symposium on Biocomputing*, 24(2019):124–135. DOI: 10.1142/9789813279827_0012. 40, 45, 83, 107

Rao, S., Ramirez, D., Braun, H., et al. (2016). An 18 kW solar array research facility for fault detection experiments, *Proc. of the 18th IEEE Mediterranean Electrotechnical Conference: Intelligent and Efficient Technologies and Services for the Citizen, (IEEE MELECON)*, pages 18–20, Limassol, Cyprus. DOI: 10.1109/melcon.2016.7495369. 93

Rao, S., Katoch, S., Narayanaswamy, V., et al. (2020). Machine learning for solar array monitoring, *Optimization, and Control*. Edited by J. Hudgins. Morgan & Claypool (Synthesis Lectures on Power Electronics). DOI: 10.2200/s01027ed1v01y202006pel013. 93

Rao, S., Spanias, A., and Tepedelenlioglu, C. (2019). Solar array fault detection using neural networks, *International Conference on Industrial Cyber-Physical Systems (ICPS)*, pages 196–200, Institute of Electrical and Electronics Engineers (IEEE), Taipei, Taiwan. DOI: 10.1109/icphys.2019.8780208. 93

Reamaroon, N., Sjoding, M. W., Lin, K., Iwashyna, T. J., and Najarian, K. (2019). Accounting for label uncertainty in machine learning for detection of acute respiratory distress syndrome, *IEEE Journal of Biomedical and Health Informatics*, 23(1):407–415. DOI: 10.1109/jbhi.2018.2810820. 84, 95

Ren, K., Yang, H., Zhao, Y., et al. (2018). A robust AUC maximization framework with simultaneous outlier detection and feature selection for positive-unlabeled classification, *IEEE Transactions on Neural Networks and Learning Systems*, pages 1–12. DOI: 10.1109/tnnls.2018.2870666. 82, 83, 84, 85, 92, 95

Ren, Y., Ji, D., and Zhang, H. (2014). Positive unlabeled learning for deceptive reviews detection, *Proc. of the Conference on Empirical Methods in Natural Language Processing (EMNLP)*, pages 488–498, Association for Computational Linguistics (ACL), Doha, Qatar. DOI: 10.3115/v1/d14-1055. 88

Reynolds, D. (2015). Gaussian mixture models, *Encyclopedia of Biometrics*, Springer. DOI: 10.1007/978-1-4899-7488-4_196. 76

Richards, F. J. (1959). A flexible growth function for empirical use, *Journal of Experimental Botany*, 10(2):290–301. DOI: 10.1093/jxb/10.2.290. 56

Rosenbaum, P. R. and Rubin, D. B. (2006). The central role of the propensity score in observational studies for causal effects, *Matched Sampling for Causal Effects*, (1083):170–184. DOI: 10.1017/CBO9780511810725.016. 24

Ruping, S. (2006). Robust probabilistic calibration, *European Conference on Machine Learning*, pages 743–750, Springer, Berlin. 9

Saito, T. and Rehmsmeier, M. (2015). The precision-recall plot is more informative than the ROC plot when evaluating binary classifiers on imbalanced datasets, *PLoS ONE*, 10(3):1–21. DOI: 10.1371/journal.pone.0118432. 40

Sansone, E., De Natale, F. G. B., and Zhou, Z. H. (2019). Efficient training for positive unlabeled learning, *IEEE Transactions on Pattern Analysis and Machine Intelligence*, 41(11):2584–2598. DOI: 10.1109/tpami.2018.2860995. 70, 81, 93, 105

Schölkopf, B., Williamson, R., Smola, A., Shawe-Taylor, J., and Piatt, J. (2000). Support vector method for novelty detection, *Advances in Neural Information Processing Systems*, pages 582–588. 28

Schroeder, M. R. and Atal, B. S. (1985). Code-excited linear prediction (CELP): High-quality speech at very low bit rates, *ICASSP*, pages 937–940. DOI: 10.1109/icassp.1985.1168147. 94

Scott, C. (2015). A rate of convergence for mixture proportion estimation, with application to learning from noisy labels, *Proc. of the 18th International Conference on Artificial Intelligence and Statistics, PMLR*, pages 838–846, JMLR and Microtome, San Diego, CA. http://arxiv.org/abs/1809.03207 33, 50

Scott, C. and Blanchard, G. (2009). Novelty detection: Unlabeled data definitely help, *International Conference on Artificial Intelligence and Statistics*, pages 464–471, PMLR, Clearwater Beach, FL. 29, 47

128   BIBLIOGRAPHY

Shanthamallu, U. S., Spanias, A., Tepedelenlioglu, C., and Stanley, M. (2017). A brief survey of machine learning methods and their sensor and IoT applications, *IEEE IISA*, Larnaca, Cyprus. DOI: 10.1109/iisa.2017.8316459. 98

Shao, Y. H., Chen, W. J., Liu, L. M., and Deng, N. Y. (2015). Laplacian unit-hyperplane learning from positive and unlabeled examples, *Information Sciences*, 314(1):152–168. DOI: 10.1016/j.ins.2015.03.066. 50

Shi, H., Pan, S., Yang, J., and Gong, C. (2018). Positive and unlabeled learning via loss decomposition and centroid estimation, *IJCAI International Joint Conference on Artificial Intelligence*, pages 2689–2695. DOI: 10.24963/ijcai.2018/373. 50, 90, 107, 108

Sickler, J. (2018). The true cost of bad reviews (and how to fix them), https://www.business2community.com/. https://www.business2community.com/crisis-management/the-true-cost-of-bad-reviews-and-how-to-fix-them-02133039 (Accessed: June 5, 2021). 88

Silver, D., Hubert, T., Schrittwieser, J., et al. (2018). A general reinforcement learning algorithm that masters chess, shogi, and Go through self-play, *Science*, 362(6419):1140–1144. DOI: 10.1126/science.aar6404. 6

Smith, A. and Elkan, C. (2004). A Bayesian network framework for reject inference, *Proc. of the International Conference on Knowledge Discovery and Data Mining (SIGKDD)*, pages 286–295, ACM, Seattle, WA. DOI: 10.1145/1014052.1014085. 89

Spanias, A. S. (1994). Speech coding: A tutorial review, *Proc. of the IEEE*, pages 1541–1582. DOI: 10.1109/5.326413. 94

Sriperumbudur, B. K., Fukumizu, K., Gretton, A., Schölkopf, B., and Lanckriet, G. R. G. (2012). On the empirical estimation of integral probability metrics, *Electronic Journal of Statistics*, 6:1550–1599. DOI: 10.1214/12-ejs722. 70

Stanley, M. and Lee, J. M. (2018). *Sensors for IoT Applications*. Edited by A. Spanias. Morgan & Claypool Publishers (Synthesis Lectures Algorithms and Software). DOI: 10.2200/S00827ED1V01201802ASE017. 98

Tanaka, D., Ikami, D., Yamasaki, T., and Aizawa, K. (2018). Joint optimization framework for learning with noisy labels, *Proc. of the IEEE Computer Society Conference on Computer Vision and Pattern Recognition*, pages 5552–5560. DOI: 10.1109/cvpr.2018.00582. 50

Tanielian, U. and Vasile, F. (2019). Relaxed sofmax for PU learning, *Proc. of the 13th ACM Conference on Recommender Systems*, pages 119–127. DOI: 10.1145/3298689.3347034. 94

Tax, D. M. J. (2002). One-class classification; Concept-learning in the absence of counter-examples. Delft University of Technology. 28

Teisseyre, P., Mielniczuk, J., and Łazecka, M. (2020). Different strategies of fitting logistic regression for positive and unlabelled data, *Lecture Notes in Computer Science (including subseries Lecture Notes in Artificial Intelligence and Lecture Notes in Bioinformatics)*, 12140 LNCS:3–17. DOI: 10.1007/978-3-030-50423-6_1. 82, 89, 92, 105, 107, 108

Tsagkatakis, G., Aidini, A., Fotiadou, K., et al. (2019). Survey of deep-learning approaches for remote sensing observation enhancement, *Sensors*, 19(18):1–39, Switzerland. DOI: 10.3390/s19183929. 30

Wang, C., Ding, C., Meraz, R. F., and Holbrook, S. R. (2006). PSoL: A positive sample only learning algorithm for finding non-coding RNA genes, *Bioinformatics*, 22(21):2590–2596. DOI: 10.1093/bioinformatics/btl441. 69, 82, 83

Ward, G., Hastie, T., Barry, S., Elith, J., and Leathwick, J. R. (2009). Presence-only data and the EM algorithm, *Biometrics*, 65(2):554–63. DOI: 10.1111/j.1541-0420.2008.01116.x. 85, 86

Warden, P. and Situnayake, D. (2020). *TinyML: Machine Learning with TensorFlow Lite on Arduino and Ultra-Low-Power Microcontrollers*, 1st ed., O'Reilly. 98

West, J. and Bergstrom, C. (no date). Which face is real. http://www.whichfaceisreal.com/ (Accessed: 5 June 2020). 8

Xu, D. and Denil, M. (2019). Positive-unlabeled reward learning. http://arxiv.org/abs/1911.00459 71

Xu, Y., Xu, Chang, Xu, Chao, and Tao, D. (2017). Multi-positive and unlabeled learning, *The International Joint Conference on Artificial Intelligence (IJCAI)*, pages 3182–3188, Melbourne, Australia. DOI: 10.24963/ijcai.2017/444. 82, 89, 91, 92, 93, 108

Yang, P., Li, X. L., Mei, J. P., Kwoh, C. K., and Ng, S. K. (2012). Positive-unlabeled learning for disease gene identification, *Bioinformatics*, 28(20):2640–2647. DOI: 10.1093/bioinformatics/bts504. 82

Yang, P., Li, X., Chua, H. N., Kwoh, C. K., and Ng, S. K. (2014). Ensemble positive unlabeled learning for disease gene identification, *PLoS ONE*, 9(5):e97079. DOI: 10.1371/journal.pone.0097079. 82

Yang, P., Liu, W., and Yang, J. (2017). Positive unlabeled learning via wrapper-based adaptive sampling, *The 26th International Joint Conference on Artificial Intelligence (IJCAI)*, pages 3273–3279, Melbourne, Australia. DOI: 10.24963/ijcai.2017/457. 82, 105, 107, 108

Yu, H. (2005). Single-class classification with mapping convergence, *Machine Learning*, 61(1–3):49–69. DOI: 10.1007/s10994-005-1122-7. 29, 31, 84, 92, 108, 109

Yu, H., Han, J., and Chang, K. C. C. (2002). PEBL: Positive example based learning for web page classification using SVM, *Proc. of the ACM SIGKDD International Conference on Knowledge Discovery and Data Mining*, pages 239–248. DOI: 10.1145/775047.775083. 61, 63, 66, 68, 92

Yu, H., Man, J., and Chang, K. C. C. (2004). PEBL: Web page classification without negative examples, *IEEE Transactions on Knowledge and Data Engineering (TKDE)*, 16(1):70–81. DOI: 10.1109/tkde.2004.1264823. 61, 63, 66, 68

Yu, S. and Li, C. (2007). PE-PUC: A graph based PU-learning approach for text classification, *Machine Learning and Data Mining in Pattern Recognition (MLDM)*, 4571:574–584. DOI: 10.1007/978-3-540-73499-4_43. 61, 62, 65, 66

Yule, G. U. (1912). On the methods of measuring association between two attributes, *Journal of the Royal Statistical Society*, 75(6):579–652. DOI: 10.2307/2340126. 39

Zadrozny, B. and Elkan, C. (2002). Transforming classifier scores into accurate multiclass probability estimates, *Proc. of the ACM SIGKDD International Conference on Knowledge Discovery and Data Mining*, pages 694–699. DOI: 10.1145/775047.775151. 9, 55

Zhang, B. and Zuo, W. (2008). Learning from positive and unlabeled examples: A survey, *Proc. of the International Symposium on Information Processing and the International Pacific Workshop on Web Mining and Web-Based Application (ISIP and WMWA)*, IEEE, Moscow, Russia. DOI: 10.1109/ISIP.2008.79. 61

Zhang, B. and Zuo, W. (2009). Reliable negative extracting based on kNN for learning from positive and unlabeled examples, *Journal of Computers*, 4(1):94–101. DOI: 10.4304/jcp.4.1.94-101. 61, 63, 66

Zhang, D. and Lee, W. S. (2005). A simple probabilistic approach to learning from positive and unlabeled examples, *Proc. of the 5th Annual UK Workshop on Computational Intelligence (UKCI)*, pages 83–87, Springer, London, England. 23, 84, 87, 92, 109

Zhang, J., Wang, Z., Yuan, J., and Tan, Y.-P. (2017). Positive and unlabeled learning for anomaly detection with multi-features, *25th ACM International Conference on Multimedia*, pages 854–862, Mountain View, CA. DOI: 10.1145/3123266.3123304. 81, 89, 90, 94, 108, 109

Zhang, J., Wang, Z., Meng, J., Tan, Y. P., and Yuan, J. (2019). Boosting positive and unlabeled learning for anomaly detection with multi-features, *IEEE Transactions on Multimedia*, 21(5):1332–1344. DOI: 10.1109/tmm.2018.2871421. 30, 61, 81, 89, 94, 108, 109

Zhang, Y., Ju, X. C., and Tian, Y. J. (2014). Nonparallel hyperplane support vector machine for PU learning, *The 10th International Conference on Natural Computation (ICNC)*, pages 703–708, IEEE, Xiamen, China. DOI: 10.1109/icnc.2014.6975922. 50

Zhao, Y., Ball, R., Mosesian, J., De Palma, J. F., and Lehman, B. (2015). Graph-based semi-supervised learning for fault detection and classification in solar photovoltaic arrays, *IEEE Transactions on Power Electronics*, 30(5):2848–2858. DOI: 10.1109/TPEL.2014.2364203. 93

Zhu, X. and Wu, X. (2004). Class noise vs. attribute noise: A quantitative study, *Artificial Intelligence Review*, 22(3):177–210. DOI: 10.1007/s10462-004-0751-8. 32, 33

Zuluaga, M. A., Hush, D., Delgado Leyton, E. J. F., Hoyos, M. H., and Orkisz, M. (2011). Learning from only positive and unlabeled data to detect lesions in vascular CT images, *Lecture Notes in Computer Science (including subseries Lecture Notes in Artificial Intelligence and Lecture Notes in Bioinformatics)*, 6893 LNCS(PART 3):9–16. DOI: 10.1007/978-3-642-23626-6_2. 84, 105

Authors' Biographies

KRISTEN JASKIE

Kristen Jaskie received her Ph.D. in Signal Processing and Machine Learning through the Electrical Engineering department and the SenSIP center at Arizona State University in Tempe, Arizona in 2021 and her B.S. and M.S. degrees in Computer Science with an emphasis in Machine Learning from the University of Washington in Seattle, Washington and the University of California San Diego in San Diego, California, respectively. Kristen is the principal ML research scientist at Prime Solutions Group and a postdoctoral researcher at ASU working with Dr. Spanias in the SenSIP center.

Kristen's research interests include machine learning and deep learning algorithm development and application, with a focus on semi-supervised learning and the positive and unlabeled learning problem. She is the author of multiple papers including "Positive and Unlabeled Learning Algorithms and Applications: a Survey," "A Modified Logistic Regression for Positive and Unlabeled Learning," and "PV Fault Detection Using Positive Unlabeled Learning."

ANDREAS SPANIAS

Andreas Spanias is a Professor in the School of Electrical, Computer, and Energy Engineering at Arizona State University (ASU). He is also the director of the Sensor Signal and Information Processing (SenSIP) center and the founder of the SenSIP industry consortium (also an NSF I/UCRC site). His research interests are in the areas of adaptive signal processing, speech processing, machine learning, and sensor systems. He and his student team developed the computer simulation software Java-DSP and its award-winning iPhone/iPad and Android versions. He is the author of two textbooks:

Audio Processing and Coding by Wiley and *DSP–An Interactive Approach* (2nd Ed.). He contributed to more than 300 papers, 11 monographs, 13 full patents, 6 provisional patents, and 10 patent pre-disclosures. He served as Associate Editor of the *IEEE Transactions on Signal Processing* and as General Co-chair of IEEE ICASSP-99. He also served as the IEEE Signal Processing Vice-President for Conferences. Andreas Spanias is co-recipient of the 2002 IEEE Donald G. Fink paper prize award and was elected Fellow of the IEEE in 2003. He served as Distinguished Lecturer for the IEEE Signal Processing society in 2004. He is a series editor for the Morgan and Claypool lecture series on algorithms and software. He recently received the 2018 IEEE Phoenix Chapter award with citation: "For significant innovations and patents in signal processing for sensor systems." He also received the 2018 IEEE Region 6 Educator Award (across 12 states) with citation: "For outstanding research and education contributions in signal processing." He was elected recently as Senior Member of the National Academy of Inventors (NAI).

Printed in the United States
by Baker & Taylor Publisher Services